Testing in the Fire Service Industry:

A Handbook for Developing Balanced and Defensible Assessments

Daniel A. Biddle, Ph.D.

Stacy L. Bell-Pilchard, M.S.

Reference: Biddle, D. A. & Bell-Pilchard, S. L. (2012). Testing in the Fire Service Industry: A Handbook for Developing Balanced and Defensible Assessments. Scottsdale, AZ: Infinity Publishing.

Fire & Police Selection, Inc. (FPSI)
193 Blue Ravine Rd., Suite 270, Folsom, California 95630
Office: (888) 990-3473 | Fax: (916) 294-4240
Email: info@FPSI.com
www.FPSI.com | www.NationalFireSelect.com

Printed in the United States of America

Published April 2012

INFINITY PUBLISHING
1094 New DeHaven Street, Suite 100
West Conshohocken, PA 19428-2713
Toll-free (877) BUY BOOK
Local Phone (610) 941-9999
Fax (610) 941-9959
Info@buybooksontheweb.com
www.buybooksontheweb.com

Contents

Overview – EEO and Testing in the Fire Service Industry

This handbook includes the key strategies your fire department will need to avoid Equal Employment Opportunity (EEO) litigation as well as keep your workforce as balanced as possible. These two goals—reducing litigation and achieving a diverse workforce—are *related* because departments that have proactive plans to build a diverse workforce also tend to be involved in fewer lawsuits. While achieving one does not guarantee the other, the strategically-managed fire department will seek them both simultaneously.

Reducing EEO litigation can only be achieved by first becoming *aware* of the basics behind EEO laws and regulations, and understanding how they work, including what sets the litigation into motion and how to avoid it. The second goal of achieving a diverse workforce can be accomplished by understanding how to properly develop and use a balanced set of assessments for hiring that will assure a well-rounded, qualified, and diverse group of employees.

This handbook includes useful information for departments to follow in creating a strategic plan for accomplishing both of these goals. The first chapter outlines the basics surrounding civil rights, Title VII of the 1991 Civil Rights Act ("1991 CRA" hereafter), and *adverse impact*, which is the main trigger for Title VII Litigation. Chapter Two provides an overview of test validation, which becomes a legal requirement whenever a testing process exhibits adverse impact. Two of the most common test validation strategies (content and criterion-related validity) are outlined in this chapter. Chapter Three highlights the importance of developing well-rounded hiring selection systems and demonstrates the very real impact that your department's testing practices can have on your bottom-line diversity results. This chapter also reviews a national study that was conducted of 130 fire executives who weighed 24 skills and abilities according to how well each contributes to success as a firefighter. Chapter Four reviews how testing practices can be *used* defensibly—whether pass/fail, banded, ranked, or weighted. Chapter Five provides step-by-step guidance on developing and implementing a content-valid, job knowledge written test. This section provides actual ratings scales to use to ensure that critical validation requirements are addressed. Chapter Six describes the key differences between using content validity and criterion-related validity evidence to demonstrate job-relatedness of a "work sample" physical ability test. This chapter also provides a detailed description on validating, administering, and scoring physical ability tests and provides sample test event descriptions. Chapter Seven includes a *validation checklist* that can be used to evaluate how likely your department's testing practices might hold up if they were targeted in a Title VII lawsuit.

Chapter 1 – The Legal Issues that Trigger Testing Litigation in the Fire Service Industry

Title VII of the 1991 Civil Rights Act

With the passage of the Civil Rights Act in 1964, it officially became illegal for employers to segregate their workforces based on race, or to otherwise consider race when making employment-related decisions. While the vast majority of employers readily adopted and welcomed this much-needed piece of civil rights legislation, some employers looked for the loopholes. One such employer was Duke Power Company which, on the very date that the Civil Rights Act became effective (July 2, 1965), established a new policy requiring applicants for jobs in the traditional white classifications (including transfers from other departments) to score sufficiently high on two aptitude tests in addition to possessing a high school diploma. This constituted three potential barricades to future career advancement. After years of deliberation in the courts regarding the validity of these tests relative to the requirements of the at-issue positions, the U.S. Supreme Court ruled (in Griggs v. Duke Power, 1971), by a vote of 8 to 0, that the tests constituted an illegal barricade to employment because they did not bear a *sufficient connection to the at-issue positions*. In other words, they were not *valid* requirements for the positions. In essence, the Court ruled that invalid practices, however neutral in intent, that cause an "adverse effect" upon a group protected by the act, are illegal.

About one year after this decision, federal civil rights enforcement agencies (*e.g.,* the U.S. Equal Employment

Opportunity Commission (EEOC) and Department of Labor's enforcement arm, the Office of Federal Contract Compliance Programs (OFCCP)) began forging versions of the federal *Uniform Guidelines* that were designed to interpret the concepts laid down in the Griggs case. Finally, in 1978, four federal agencies solidified a set of *Uniform Guidelines* that still serve to interpret the "job relatedness" and "validity" requirements from the Griggs case, as reported by the Questions & Answers accompanying the Guidelines:

> Question: What is the basic principle of the Guidelines? Answer: A selection process which has an adverse impact on the employment opportunities of members of a race, color, religion, sex, or national origin group and thus disproportionately screens them out is unlawfully discriminatory unless the process or its component procedures have been validated in accord with the Guidelines, or the user otherwise justifies them in accord with Federal law… This principle was adopted by the Supreme Court unanimously in Griggs v. Duke Power Co., 401 U.S. 424, and was ratified and endorsed by the Congress when it passed the Equal Employment Opportunity Act of 1972, which amended Title VII of the Civil Rights Act of 1964 (*Uniform Guidelines*, Q&A #2).

However, the principles laid down in Griggs have not gone without challenge. In 1989, the U.S. Supreme Court handed down a decision that drastically changed the Griggs principles that had been incorporated into the *Uniform Guidelines*. This landmark case, Ward's Cove Packing Co. v.

Atonio (1989), revised the legal burdens surrounding proof of discrimination from the Griggs-based employer's burden (where the employer had to prove that any disparate impact[1] caused by the testing practices was justified by job relatedness/validity) to the burden of proof *remaining with the plaintiff at all times*. Under Ward's Cove, all the employer was required to do was "produce a business justification" that would justify the disparate impact—a legal burden much easier to address than the "job-relatedness/business necessity" standard laid down in Griggs.

In 1991, an Act of Congress *reversed* Wards Cove (and several other similar U.S. Supreme Court decisions it found disagreeable). This Act—the 1991 Civil Rights Act (or "1991 CRA")—shifted the validation burden back to the Griggs standard by specifically stating the two conditions under which an employer's testing practice will be deemed illegal:

> A(i) a complaining party demonstrates that a respondent uses a particular employment practice that causes a disparate impact on the basis of race, color, religion, sex, or national origin, and the respondent fails to demonstrate that the challenged practice is job-related for the position in question and consistent with business necessity; OR, A(ii) the complaining party makes the demonstration described in subparagraph (C) with respect to an alternate employment practice, and the respondent refuses to adopt such alternative employment practice (Section 2000e-2[k][1][A][i]).

The above section from the 1991 CRA effectively summarizes the current "law of the land" with respect to disparate impact and test validation. Notice that the language is *job-specific*: "the challenged practice is job-related *for the position in question* and consistent with business necessity." Thus, by definition, evaluating the validity of a test is a test-by-test, job-by-job determination. Also, note that Section A(ii) is preceded by the word OR, indicating that there are clearly two routes available for proving adverse impact, with the first being the "classic" method (adverse impact with no validity), and the second being the showing of a "alternative employment practice" that could be used with less adverse impact. While cases have been tried using this alternative strategy, this concept will not be discussed further here in order to focus on the classic type which is the most common.

Legal Update: What Has Happened Since the 1991 CRA?

Since the passage of the 1991 CRA, there has been a steady stream of disparate impact cases in the fire service industry. As of the time of this writing, no cases have changed the foundational framework of the Griggs standard codified in the 1991 CRA. There have, however, been some changes on the legal front regarding the extent to which race-conscious actions can be taken *after* a testing process has been administered.

These changes were spurred by the Ricci v. DeStefano case tried in the U.S. Supreme Court in 2009. The Ricci case arose from a lawsuit brought against the City of New Haven, Connecticut, by 19 City firefighters who claimed that the City discriminated against them with regard to promotions. The firefighters (17 whites and 2 Hispanics) had all passed the promotional tests for the Captain or Lieutenant positions but, just prior to finalizing their promotion eligibility, the City invalidated the test results because none of the black firefighters who passed the exam had scored high enough to be considered for the positions. The City claimed that their reason for cancelling the list was that they "feared a lawsuit over the test's disparate impact."

Ultimately, the Supreme Court found this decision to be discriminatory because the City lacked the *strong basis in evidence* that it would have lost a disparate impact lawsuit (because the tests were not sufficiently valid). While there was no real contention regarding the existence of adverse impact, the evidence to weigh the validity of the test was never admitted into evidence. Rather, the City blocked the validation evidence (by requesting that the test vendor not send them a validation report—one which was already scheduled for delivery) and the plaintiffs (whites in this case) also did not want to contend the validity of the case.

In this interesting set of circumstances, the actual validity of the tests was in a "pickle" situation—the City was motivated to prove that the tests *were not* valid (because they were trying to claim that their actions of redacting the exam results was justified because the tests were "sufficiently invalid") and the plaintiffs also did not want the validity contested (because they wanted the exam results to stand and not be invalidated). Based on this very unique set of circumstances, a split (5-4) decision was rendered by the Court on June 29, 2009. The Court's ruling held that the City's decision to ignore the test results violated Title VII of the Civil Rights Act of 1991.

Immediately after the case, legal blogs and presentations spread through the Internet—with some even claiming that the Griggs/1991 CRA legal foundation had been changed (which was clearly not the case). In fact, to bring clarity to the distinctions between the Ricci case—which was essentially a *disparate treatment case that involved a test*—some courts began clarifying that the Griggs/1991 CRA standard was still alive and well. One such case was Vulcan Society v. City of New York (2009) which was decided just weeks after the Ricci ruling. In Vulcan, the judge clarified the distinctions between the Ricci ruling (disparate *treatment*) and typical disparate *impact* cases:

> Before proceeding to the legal analysis, I offer a brief word about the Supreme Court's recent decision in Ricci … I reference Ricci not because the Supreme Court's ruling controls the outcome in this case; to the contrary, I mention Ricci precisely to point out that it does not. In Ricci, the City of New Haven had set aside the results of a promotional examination, and the Supreme Court confronted the narrow issue of whether New Haven could defend a violation of Title VII's disparate treatment provision by asserting that its challenged employment action was an attempt to comply with Title VII's disparate impact provision. The Court held that such a defense is only available when "the employer can demonstrate a strong basis in evidence that, had it not taken the action, it would have been liable under the disparate-impact statute" (Id. at 2664). In contrast, this

> case presents the entirely separate question of whether Plaintiffs have shown that the City's use of [the Exams] has actually had a disparate impact upon black and Hispanic applicants for positions as entry-level firefighters. Ricci did not confront that issue…The relevant teaching of Ricci, in this regard, is that the process of designing employment examinations is complex, requiring consultation with experts and careful consideration of accepted testing standards. As discussed below, these requirements are reflected in federal regulations and existing Second Circuit precedent. This legal authority sets forth a simple principle: municipalities must take adequate measures to ensure that their civil service examinations reliably test the relevant knowledge, skills and abilities that will determine which applicants will best perform their specific public duties.

The gears of several disparate impact cases continued to turn during and after the Ricci case, with the Griggs standards in full effect. Should a case come along some day that does change the Griggs standard, a commensurate change to the *Uniform Guidelines* will also be likely.

Adverse Impact: The Main Trigger for Title VII Litigation

Each situation where a Title VII claim is made has a common denominator: adverse impact. The way the federal law currently stands, plaintiffs cannot even bring a lawsuit unless and until a test causes adverse impact. Because this trigger underlies all of EEO-related litigation (at least of the disparate impact variety), some attention will now be turned to defining "adverse impact."

Rather than entering into a protracted academic definition of adverse impact, the topic will be simplified into just a few paragraphs.[2] When defining adverse impact, there are some important, key terms relevant to the matter. These include the 80% test, statistical significance, and practical significance. Each is described briefly below.

80% Test

This test is calculated by dividing the focal group's (the focal group is typically minorities or women) passing rate on a test by the reference group's (typically whites or men) passing rate. Any resulting value less than 80% constitutes a "violation" of this test. This test was originally framed in 1972 (see Biddle, 2011), was codified in the *Uniform Guidelines* in 1978, and has been referenced in hundreds of court cases. Despite its widespread use, it should not be regarded as the final litmus test for determining adverse impact. That position is held exclusively by statistical significance tests, described next.

Statistical Significance Tests

A "statistically significant finding" is one that raises the eyebrows of the researcher and causes him to think, "I have found something here, and it is not likely due to chance." So, in the realm of an adverse impact analysis, if a researcher conducts an adverse impact analysis and obtains a statistically significant result, they can state that a legitimate trend, and not a chance relationship, actually exists (with a reasonable level of certainty).

Statistical significance tests result in a p-value ("p" for probability), with p-values ranging from 0 to +1. A p-value of 0.01 means that the odds of the event occurring by chance is only 1%. A p-value of 1.0 means that there is essentially a 100% certainty that the event is "merely a chance occurrence," and cannot be considered as a "meaningful finding." P-values of .05 or less are said to be "statistically significant" in the realm of EEO analyses. This .05 level (or 5%) corresponds with the odds ratio of "1 chance in 20." This 5% chance level is the p-value threshold that has been endorsed in nearly every adverse impact case or federal enforcement setting.

Conducting a statistical significance adverse impact analysis is very straightforward, provided that a statistical software program is used.[3] The process is completed by applying a statistical test to a 2 X 2 table, where the success rates (*e.g.*, pass or failing a test) of two groups (*e.g.*, Whites and Asians) are compared. See the example below.

2 X 2 Table Example

Group	Promoted	Not Promoted
Whites	30	20
Asians	20	30

There are over 20 possible statistical tests that can be used for computing the statistical significance of a 2 X 2 table, including "estimation" and "exact" methods, and various models that make certain assumptions regarding how the 2 X 2 table itself is constructed (see Biddle & Morris, 2011 for a complete discussion). For example, using a "estimation" technique (such as the Chi-Square computation) on the table above returns a p-value of .046 (below the .05 level needed for a "statistically significant" finding); whereas using a more exact method (the Fisher's Exact Test with Lancaster's Mid-P Correction)[4] returns a p-value of .06 (not significant).

Practical Significance

The concept of practical significance in the EEO analysis field was first introduced by Section 4D of the *Uniform Guidelines* ("Smaller differences in selection rate may nevertheless constitute adverse impact, where they are significant in *both statistical and practical terms* ..."). Practical significance tests are applied to adverse impact analyses to evaluate the "practical impact" (typically reported as the shortfall pertaining to the group with the lower passing rate) or "stability" of the results (evaluating whether a statistically significant finding still exists after changing the passing/failing numbers of the disadvantaged group). While this concept enjoyed a run in the federal court system (Biddle, 2011), it has more recently been met with a considerable level of disagreement and condemnation in the courts.[5] For example, in the most recent circuit-level case dealing with practical significance, the court stated:

> Similarly, this Court has never established "practical significance" as an independent

> requirement for a plaintiff's prima facie disparate impact case, and we decline to do so here. The EEOC Guidelines themselves do not set out "practical" significance as an independent requirement, and we find that in a case in which the statistical significance of some set of results is clear, there is no need to probe for additional "practical" significance. Statistical significance is relevant because it allows a fact-finder to be confident that the relationship between some rule or policy and some set of disparate impact results was not the product of chance. This goes to the plaintiff's burden of introducing statistical evidence that is "sufficiently substantial" to raise "an inference of causation." Watson, 487 U.S. at 994-95. There is no additional requirement that the disparate impact caused be above some threshold level of practical significance. Accordingly, the District Court erred in ruling "in the alternative" that the absence of practical significance was fatal to Plaintiffs' case (Stagi v. National Railroad Passenger Corporation, 2010).

For these reasons, while the *Uniform Guidelines* are clear that practical significance evaluations (such as shortfalls and statistical significance "stability" tests) are *conceptually relevant* to adverse impact analyses, employers should "tread carefully" when evaluating the practical significance of adverse impact analysis results. While the concept is relevant, it will ultimately be left to a judge to decide whether (and to what extent) practical significance can be used in court. Certainly it would be a risky endeavor to adopt hard-and-fast practical significance rules when analyzing adverse impact.

Chapter 2 – Validation: The Legal Requirements When Tests Exhibit Adverse Impact

Test validation is often misunderstood. Yes, validation is a legal obligation whenever an employer's test exhibits adverse impact. But it is much more than that—validation is actually a scientific process for ensuring that your department's testing process is focusing on the *key competencies that are needed for job success*. Without a properly-validated hiring process, hiring authorities might as well pull names out of a hat! So validation should be regarded as having two benefits: (1) utility (the benefits enjoyed by employers by hiring highly-qualified candidates), and (2) legal defensibility.

Validation Techniques and Legal Requirements for Testing

The two validation strategies that are applied most frequently in practice are content validity and criterion-related validity. Both strategies are supported by the *Uniform Guidelines*, professional standards, and numerous court cases. While there are several possible angles to develop tests under either strategy, the most basic components of each are discussed below.

Content Validity

Content validity evidence is gathered by demonstrating a nexus (*i.e.,* a connection) between the test and important job requirements. If conducted from the ground-up, a typical content validation study will include the following steps:

1. **Conducting Job Analysis Research.** Establishing content validity evidence requires having a clear understanding of what the job requires—especially the areas that are targeted by the test. Generally speaking, content validity evidence is stronger in circumstances where a clear picture of the job has been developed through a thorough job analysis process.

2. **Developing a Clear Test Plan.** A Test Plan identifies the key knowledges, skills, abilities, and personal characteristics (KSAPCs) identified by the job analysis process as being necessary on the *first day* of employment. Ideally, the most important KSAPCs are included in the Test Plan.

3. **Connecting the Test to the Job.** A process needs to be completed that *establishes or demonstrates* a clear nexus between the test and the important job KSAPCs. This can be completed by using the expert opinions of either methodology (testing) experts or Job Experts. Ultimately, the KSAPCs measured by the test need to be linked to the important KSAPCs of the job, which are then linked to the important job duties. This three-way process establishes content validity.

4. **Establishing How the Tests will be *Used*.** Adopting a content validity process requires *using* (*e.g.*, ranking, pass/fail, banded) the test results in a way that accurately reflects how the important KSAPCs measured by the test are *actually applied on the job*. For example, possessing basic math skills is necessary for being a competent firefighter, but possessing increasingly higher

levels of this skill does not necessarily translate into superior performance as a firefighter. Other skills, such as teamwork and interpersonal skills, are more likely to *differentiate performance* between firefighters when held at above-minimum levels. Following in this same spirit, tests measuring basic math should be used on a pass/fail (cutoff) basis, whereas tests measuring differentiating KSAPCs should be the primary basis for ranking decisions. A complete discussion of test use considerations is provided in Chapter Four.

Criterion-related Validity

Criterion-related validity is statistical in nature, and is established by demonstrating a significant correlation between the test and some important aspect of job performance. For example, a department might have the supervisory staff assign job performance ratings to the firefighters, run the firefighters through a physical ability test, then conduct a statistical correlation between the test and job performance ratings to assess whether they are significantly correlated.

Criterion-related validity studies can be conducted in one of two ways: using a predictive model or a concurrent model. A predictive model is conducted when applicant test scores are correlated to subsequent measures of job performance (*e.g.,* six months after the tested applicants are hired). A concurrent model is conducted by giving a test to incumbents who are currently on the job and then correlating these scores to current measures of job performance (*e.g.,* performance review scores, supervisor ratings, etc.). The following steps can be completed to conduct a predictive criterion-related validity study:

1. Be sure that your department has at least 150 subjects to include in the study (applicants who will take the pre-employment tests and will be subsequently hired). This minimum sample size is recommended because, in most situations, it provides adequate statistical power for detecting a meaningful correlation (e.g., a sample size of 134 provides a 90% likelihood of finding a .25 correlation, if such a correlation exists in the target population).

2. Conduct a job analysis (see previous section) or a "review of job information" to determine the important aspects of the job that should be included in the study (both from the testing side and the rating side).

3. Develop one or more criterion measures by developing subjective (*e.g.,* job performance rating scales) or objective measures (*e.g.,* absenteeism, work output levels) of critical areas from the job analysis or job information review. A subjectively-rated criterion should only consist of performance on a *job duty* (or group of duties). In most cases, it should not consist of a supervisor/co-worker rating on the incumbent's *level of KSAPCs* (a requirement based on Section 15B5 of the *Uniform Guidelines*) unless the KSAPCs are clearly linked to observable work behaviors, or they are sufficiently operationally defined. It is important that these measures have sufficiently high reliability (at least .60 or higher is preferred).

4. Work with Job Experts and supervisors, trainers, other management staff, and the job analysis data to form

solid speculations ("hypotheses") regarding which KSAPCs "really make a difference" in the high/low scores of such job performance measures (above).

5. Develop tests that are reliable measures of those KSAPCs. Choosing tests that have reliability of .70 or higher is preferred.[6]

6. After a period of time has passed and criterion data has been gathered (*e.g.*, typically between 3 and 12 months), correlate each of the tests to the criterion measures using the =PEARSON command in Microsoft Excel and evaluate the results.

To complete a concurrent criterion-related validation study, complete steps 2–5 above and replace step 6 by administering the test to the current incumbent population and correlate the test scores to current measures of job performance.

In either study design, the resulting correlation coefficient must, at a minimum, be statistically significant at the .05 level (before making any corrections). Ideally, it should also be sufficiently strong to result in practical usefulness in the hiring process. The U.S. Department of Labor (2000, pp. 3-10) has provided reasonable guidelines for interpreting correlation coefficients, with coefficients between .21 and .35 classified as "likely to be useful" and coefficients higher than .35 as "very beneficial."

Chapter 3 – Building a Balanced Hiring Program by Sorting the Key Competencies for Success

Sometimes fire executives ask, "Which entry-level firefighter test is the best?" The real question is, "What is the best *set* of tests for selecting the most qualified, well-rounded group of firefighters?" Until the testing field advances to using ultra-precision testing instruments that can conduct "deep scans" of each applicant's true potential as a firefighter in just 20 minutes, the testing process for screening entry-level applicants will need to consist of a *wide range* of tests that measure a wide range of abilities. It will also continue to be an arduous process—by the time each applicant's resume, background, written, and interview screens have been reviewed, tallied, and scored, the time investment typically exceeds several hours per candidate.

While the process will always be ardurous, carefully choosing which tests to use and which competency areas to measure are the most important strategic factors to consider. To explore options for these key decision factors, this section provides the results of a national survey that was completed by 130 fire executives (fire chiefs of all levels) that asked them to identify the most important competency areas needed for the entry-level firefighter position. Specifically, they were asked to assign 100 points to three major competency areas[7]:

- Cognitive/academic (such as reading, math, writing);
- Personal characteristics (such as working under stress, allegiance, integrity); and,

- Physical abilities (such as upper body strength, stamina, speed).

After the survey participants assigned 100 points to these three major competency areas, they were asked to assign 100 points *within* each. The results are shown in Table 1.

Table 1. Entry-Level Firefighter Competency Weights

Cognitive/Academic (32% of Total)	% Importance
Math	10%
Reading	14%
Verbal Communication	15%
Writing	12%
Map Reading	8%
Problem Solving	15%
Strategic Decision-Making	13%
Mechanical Ability	12%
Personal Characteristics (40% of Total)	**% Importance**
Teamwork	12%
Working Under Stress	10%
Allegiance/Loyalty	9%
Truthfulness/Integrity	13%
Public Relations	8%
Emotional Stability	10%
Sensitivity	8%
Proactive/Goal-Oriented	8%
Thoroughness/Attention to Detail	9%
Following Orders	10%
Physical Abilities (28% of Total)	**% Importance**
Wrist/Forearm Strength	13%
Upper Body Strength	17%
Lower Torso and Leg Strength	17%
Speed	12%
Dexterity, Balance, and Coordination	16%
Endurance	21%

The survey revealed that supervisory fire personnel valued the cognitive/academic domain as 32% of the firefighter position, personal characteristics as 40%, and physical as 28%. It should be pointed out that most professionals who have seasoned their careers in the fire service would be the first to admit that the typical entry-level testing process does not reflect these ratios. In fact, most testing processes focus mostly on the cognitive/academic areas (typically through a pass/fail written test), use a basic physical ability test (again, pass/fail), and only measure a very limited degree of personal characteristics through an interview process (but only for the final few candidates who are competing for a small number of open positions).

Not only does this result in a disconnect between the competencies that are required for the job and those that are included in the screening process, it results in hiring processes that can unnecessarily cause adverse impact against minorities and/or women. For example, a hiring process that focuses exclusively on cognitive/academic skills will magnify adverse impact against minorities. The other cost for using such an unbalanced hiring process is that it leaves a massive vacuum (40%, to be precise) in the personal characteristics area—leaving these important competencies completely untapped. A hiring process that over-emphasizes the importance of physical abilities will amplify adverse impact against women.

Beyond the adverse impact issues, other substantial problems occur when a fire department adopts an unbalanced hiring process. A testing process that leaves out (or under-measures) important cognitive/academic skills will likely result in washing out a significant number of cadets in the academy, and can also lead to poor performance on the job. On the other hand, over-measuring this area while under-measuring personal characteristics, could lead to a group of book-smart firefighters who have no idea how to work cooperatively in the close living conditions required of firefighters. Under-measuring physical abilities in the hiring process while over-measuring personal characteristics could result in a group of firefighters who cannot perform the strenuous physical requirements of the job—especially as they age in the fire service. Clearly, it is important in the testing process to find a balanced approach.

Challenges to Building a Balanced Testing Program and Recommendations for Success

The most significant challenge to building an effective testing program for entry-level firefighters lies with testing interpersonal skills (personal characteristics). This is because these skills—such as teamwork and interpersonal skills—are crucial ingredients for success but they are the most difficult to measure in a typical testing format. For example, developing a math test is easy; developing a test for measuring teamwork skills is not—but the latter was rated as more important for overall job success!

The reason for this is that many skills and abilities are "concrete" (opposed to theoretical and abstract). An applicant's math skills can be readily tapped using questions that measure numerical skills at the same level these skills are required on the job. Abstract or soft skills like teamwork are more difficult to measure during a one- or two-hour testing session.

Fortunately, there are some effective and innovative testing solutions available. Tables 2-4 provide suggested test methodologies and tools for each of the key competencies needed for being a fully-rounded firefighter.

Table 2. Proposed Solutions for Testing Cognitive/Academic Competencies for Entry-Level Firefighters

Cognitive/Academic (32% Overall Importance)	Weight	Proposed Testing Solution	Typical Validation Method
Math	10%	Use written or "work sample" format; measure using a limited number of multiple-choice items. Balance various types of math skills (add/subtract/multiple/divide, etc.).	CV
Reading	14%	Measure using either (1) "Test Preparation Manual" approach (where the applicants are given a manual and asked to study it for a few weeks prior to taking the test based on the Manual; or (2) a short reading passage containing material at a similar difficulty/context to the job that applicants are allowed to study during the testing session and use to answer related test items.	CV
Verbal Communication	15%	While a Structured Interview is the best tool for measuring this skill (because the skill includes verbal and non-verbal aspects), some level of this skill can be measured using word recognition lists or sentence clarity items.	CV
Writing	12%	Measure using writing passages or word recognition lists, sentence clarity, and/or grammar evaluation items.	CV
Map Reading	8%	Measure using maps and related questions asking applicants how they would maneuver to certain locations. Include directional awareness.	CV
Problem Solving	15%	Measure using word problems that measure reasoning skills in job-rich contexts.	CRV
Strategic Decision-Making	13%	While a Structured Interview is the best tool for measuring this skill (because the applicant can be asked to apply this skill in firefighter-specific scenarios), some level of this skill can be measured using word problems or other contexts supplied in written format where applicants can consider cause/effect of certain actions.	CV
Mechanical Ability	12%	Using CV, measure mechanical comprehension skills such as leverage, force, and mechanical/physics contexts regarding weights, shapes, and distances. Also can measure spatial reasoning (when using a CRV validation strategy).	CV/CRV

Notes: CV: Content Validity; CRV: Criterion-related validity.

Table 3. Proposed Solutions for Personal Characteristics for Entry-Level Firefighters

Personal Characteristics (40% Overall)	Weight	Proposed Testing Solution	Typical Validation Method
Teamwork	12%	Under a CV strategy, a Situational Judgment Test (SJT) can be used for measuring these skills. Alternatively, a custom personality test can be developed using CRV. While these types of assessments can measure *whether an applicant knows* the most appropriate response (using an SJT) or has the best attitude or disposition (personality test), they are limited in that they cannot measure whether an applicant *would actually respond* in such a way. For these reasons, measuring the *underlying traits* that tend to generate these positive behaviors is typically the most effective strategy. Structured Interviews can also provide useful insight into these types of competencies, as well as background and reference evaluations. However, these tools are time consuming and expensive, so measuring these areas in the testing stage is an effective strategy.	CV/CRV
Working Under Stress	10%		
Allegiance/Loyalty	9%		
Truthfulness/Integrity	13%		
Public Relations	8%		
Emotional Stability	10%		
Sensitivity	8%		
Proactive/Goal-Oriented	8%	These competencies can be effectively measured using either an SJT (using a CV or CRV strategy) or a Conscientiousness (CS) scale (using a CRV strategy). A CS test can be developed using just 20-30 items (using likert-type responses). Such tests are typically successful in predicting job performance in fire settings.	CV/CRV
Thoroughness/ Attention to Detail	9%		
Following Orders	10%		

Notes: CV: Content Validity; CRV: Criterion-related validity.

Table 4. Proposed Solutions for Testing Physical Abilities for Entry-Level Firefighters

Physical Abilities (28% Overall)	Weight	Proposed Testing Solution	Typical Validation Method
Wrist/Forearm Strength	13%	It is the opinion of the authors that these unique physical competencies should be *collectively and representatively* measured in a work-sample style Physical Ability Test (PATs) (using a content validity strategy). While other types of tests (such as clinical strength tests) that do not directly mirror the requirements of the job can be used (if they are based on CRV), using a high-fidelity work sample typically has greater benefits.	CV
Upper Body Strength	17%		CV
Lower Torso/Leg Strength	17%		CV
Dexterity, Balance, Coordination	16%		CV
Speed	12%	Strenuous work-sample PATs can measure some level of endurance (and speed) if they are continuously-timed and exceed at least five minutes in length. Actual cardiovascular endurance levels can only be measured using a post-job-offer V02 maximum test (which would require using a CRV strategy) (see Chapter Six).	CV/CRV
Endurance	21%		

Notes: CV: Content Validity; CRV: Criterion-related validity.

The importance weights displayed in the tables above may or may not be representative of individual fire departments. For example, when designing a Physical Ability Test, some departments serve communities that have more multiple structure or high-rise fires than others, some have a higher occurrence rate of EMS incidents, etc. Therefore, we recommend that each fire department investigate the relative importance of these various competencies and the tests used to measure them (discussed further in the next section).

Chapter 4 – Test Use: Setting Pass/Fail Cutoffs, Banding, Ranking, and Weighting

Because validation has to do with the *interpretation of scores*, a perfectly valid test can be *invalidated* through improper use of the scores. Conducting a search through professional testing guidelines (*e.g.,* the *Principles*, 2003, and *Standards*, 1999), the *Uniform Guidelines* (1978), and the courts, one can find an abundance of instruction surrounding how test scores should be used. The safe way to address this complex maze of guidelines is to be sure that tests are *used in the manner that the validation evidence supports.*

If classifying applicants into two groups—qualified and unqualified—is the end goal, the test should be used on a pass/fail basis (*i.e.,* an absolute classification based on achieving a certain level on the test). If the objective is to make relative distinctions between substantially equally qualified applicants, then banding is the approach that should be used. Ranking should be used if the goal is making decisions on an applicant-by-applicant basis (making sure that the requirements for ranking discussed herein are addressed). If an overall picture of each applicant's combined mix of competencies is desired, then a weighted and combined selection process should be used.

For each of these options, different types of validation evidence should be gathered to justify the corresponding manner in which the scores will be used and interpreted. This section explains the steps that can be taken to develop and justify each.

Developing Valid Cutoff Scores

Few things can be as frustrating as being the applicant who scored 69.9% on a test with a 70% cutoff! Actually, there *is* one thing worse: finding out that the employer elected to use 70% as a cutoff for *no good reason whatsoever*. Sometimes this arbitrary cutoff is chosen because it just *seems* like a "good, fair place to set the cutoff" or because 70% represents a C grade in school. Arbitrary cutoffs simply do not make sense, either academically or practically. Further, they can incense applicants who might come to realize that a meaningless standard in the selection process has been used to make very *meaningful decisions* about their lives and careers.

For these reasons, and because the federal courts have so frequently rejected arbitrary cutoffs that have adverse impact, it is essential that practitioners use *best practices* when developing cutoffs. And, when it comes to best practices for developing cutoffs, there is perhaps none better than the *modified Angoff method.*[8] The Angoff method makes good practical sense, Job Experts can readily understand it, applicants can be convinced of its validity, the courts have regularly endorsed it,[9] and it stands up to academic scrutiny.

Developing a cutoff score using this method is relatively simple: Job Experts review each item on a written test and provide their "best estimate" on the percentage of minimally qualified applicants they believe would answer the item correctly (*i.e.,* each item is assigned a percentage value). These ratings are trimmed and then averaged[10] and a valid cutoff for the test can be developed. The *modified* Angoff method adds a slight variation: After the test has been administered, the cutoff level set using the method above is lowered by 1, 2, or 3

Conditional Standard Errors of Measurement (C-SEMs)[11] to adjust for the unreliability of the test.

The *Uniform Guidelines* require that pass/fail cutoffs should be ". . . set so as to be reasonable and consistent with the normal expectations of acceptable proficiency in the workforce" (Section 5H). The modified Angoff method addresses this requirement on an item-by-item basis.

Setting Cutoffs that are Higher than the "Minimum Cutoff Level"

It is not uncommon for large fire departments to be faced with situations where thousands of applicants apply for only a handful of open positions. What should be done if the department cannot feasibly process all applicants who pass the validated cutoff score? Theoretically speaking, all applicants who pass the modified Angoff cutoff are qualified; however, if the department simply cannot process the number of applicants who pass the given cutoff, two options are available.

The first option is to use a cutoff that is *higher* than the minimum level set by the modified Angoff process. If this option is used, the *Uniform Guidelines* are clear that the degree of adverse impact should be considered (see Section 3B and 5H). One method for setting a higher cutoff is to subtract one Standard Error of Difference (SED)[12] from the highest score in the distribution, then passing all applicants in this score band. Using the SED in this process helps ensure that all applicants within the band are *substantially equally qualified.* Additional bands can be created by subtracting one SED from the score immediately below the band for the next group, and repeating this process until the first cutoff score option is reached (*i.e.*, one Conditional SEM below the cutoff score). This represents the distinguishing line between the qualified and unqualified applicants.

While this option may be useful for obtaining a smaller group of applicants who pass the cutoff score and are substantially equally qualified, a second option is strict rank ordering. Strict rank ordering is not typically advised on written tests because of the high levels of adverse impact that are likely to result and because written tests typically only include a narrow measurement of the wide competency set that is needed for job success. To hire or promote applicants in strict rank order on a score list, the employer should be careful to ensure that the criteria in the Ranking section below are sufficiently addressed.

Banding

In some circumstances, applicants are rank-ordered on a test and hiring decisions between applicants are based upon score differences at the one-hundredth or one-thousandth decimal place (*e.g.,* applicant A who scored 89.189 is hired before applicant B who scored 89.188, etc.). The troubling issue with this practice is that, if the test were administered a second time, applicants A and B could very likely change places! In fact, if the reliability of the test was low and the standard deviation was large, these two applicants could be separated by several whole points.

Banding addresses this issue by using the Standard Error of Difference (SED) to group applicants into "substantially equally qualified" score bands. The SED is a tool that can be used by practitioners for setting a confidence interval around scores that are substantially equal. Viewed another way, it can be used for determining scores in a distribution that represents *meaningfully different* levels of the competencies measured by the test. The technical procedures for computing score bands using the C-SEM and SED are outlined in Biddle (2011).

Banding has been a hotly debated issue in the personnel field, especially over the last 20 years.[13] Proponents of strict rank ordering argue that making hiring decisions in rank order preserves meritocracy and ultimately ensures a slightly more qualified workforce. Supporters of banding argue that, because tests cannot adequately distinguish between small score differences, practitioners should remain blind to miniscule score differences between applicants who are within the same band. They also argue that the practice of banding will almost always produce less adverse impact than strict rank ordering.[14] While these two perspectives may differ, various types of score banding procedures have been successfully litigated and supported in court,[15] with the one exception being the decision to band *after* a test has been administered, if the *only* reason for banding was to reduce adverse impact (Ricci, 2009). Thus, banding remains as an effective tool that can be used in most personnel situations.

Ranking

The idea of hiring applicants in strict order from the top of the list to the last applicant above the cutoff score is a practice that has roots back to the origins of the merit-based civil service system. The limitation with ranking, as discussed above, is that the practice treats applicants who have almost tied scores as if their scores are meaningfully different *when we know that they are not*. The C-SEM shows the degree to which scores would likely shuffle if the test was hypothetically administered a second time.

Because of these limitations, the *Uniform Guidelines* and the courts have presented rather strict requirements surrounding the practice of rank ordering. These requirements are provided below, along with some specific recommendations on the criteria to consider before using a test to rank order applicants.

Section 14C9 of the *Uniform Guidelines* states:

> If a user can show, by a job analysis or otherwise, that a higher score on a content valid test is likely to result in better job performance, the results may be used to rank persons who score above minimum levels. Where a test supported solely or primarily by content validity is used to rank job candidates, the test should *measure those aspects of performance which differentiate among levels of job performance*.

Performance differentiating KSAPCs distinguish between acceptable and above-acceptable performance on the job. Differentiating KSAPCs can be identified either *absolutely* (each KSAPC irrespective of the others) or *relatively* (each KSAPC relative to the others) using a "Best Worker" likert-type rating scale to rate KSAPCs regarding the extent to which it distinguishes the "minimal" from the "best" worker. KSAPCs that are rated high on the Best Worker rating are those that, when performed above the "bare minimum," distinguish the "best" performers from the "minimal."

For example, possessing basic math skills is a necessity for being a competent firefighter, but possessing increasingly higher levels of this skill does not necessarily translate into superior performance as a firefighter. Other skills, such as teamwork and interpersonal skills, are more likely to differentiate performance when held at above-minimum levels.

A strict rank ordering process should not be used on a test that measures KSAPCs that are only needed at *minimum levels* on the job and do not distinguish between acceptable and above-acceptable job performance (see the *Uniform Guidelines* Questions & Answers #62). Content validity evidence to support ranking can be established by linking the parts of a test to KSAPCs that are performance differentiating.[16] So, if a test is linked to a KSAPC that is "performance differentiating" either *absolutely* or *relatively* (*e.g.,* with an average differentiating rating that is 1.0 standard deviation above the average rating compared to all other KSAPCs), some support is provided for using the test as a ranking device.

While the Best Worker rating provides some support for using a test as a ranking device, there are additional factors to consider before making a decision to use a test in a strict rank-ordered fashion:

1. Is there adequate score dispersion in the distribution (or a "wide variance of scores")? Rank ordering is usually not preferred if the applicant scores are "tightly bunched together"[17] because such scores are "tied" to even a greater extent than if they were more evenly distributed. One way to evaluate the dispersion of scores is to use the C-SEM to evaluate if the score dispersion is adequately spread out within the relevant range of scores when compared to other parts of the score distribution. For example, if the C-SEM is very small (*e.g.,* 2.0) in

the range of scores where the strict rank ordering will occur (*e.g.*, 95 – 100), but is very broad throughout the other parts of the score distribution (*e.g.*, double or triple the size), the score dispersion in the relevant range of interest (*e.g.*, 95-100) may not be sufficiently high to justify rank ordering.

2. Does the test have high reliability? Typically, reliability coefficients should be .85 to .90 or higher for using the results in strict rank order.[18] If a test is not reliable (or "consistent") enough to "split apart" candidates based upon very small score differences, it should not be used in such a way that considers small differences between candidates as meaningful.

While the guidelines above should be considered when choosing a rank ordering or pass/fail strategy for a test, the extent to which the test measures KSAPCs[19] that are performance differentiating should be the *primary consideration.*

Employers using a test that is based on criterion-related validity evidence have more flexibility to use ranking than with tests based on content validity. This is because criterion-related validity demonstrates scientifically what content validity can only speculate is occurring between the test and job performance. Criterion-related validity provides a correlation coefficient that represents the strength or degree of correlation relationship between some aspects of job performance and the test.

While the courts have regularly endorsed criterion-related validity studies, they have placed some minimum thresholds for the correlation value necessary (typically .30 or higher) for strict rank ordering on a firefighter tests based on criterion-related validity:

- Brunet v. City of Columbus (1993). This case involved an entry-level firefighter Physical Capacities Test (PCT) that had adverse impact against women. The court stated, "The correlation coefficient for the overall PCT is .29. Other courts have found such correlation coefficients to be predictive of job performance, thus indicating the appropriateness of ranking where the correlation coefficient value is .30 or better."

- Boston Chapter, NAACP Inc. v. Beecher (1974). This case involved an entry-level firefighter written test. Regarding the correlation values, the court stated, "The objective portion of the study produced several correlations that were statistically significant (likely to occur by chance in fewer than five of one hundred similar cases) and practically significant (correlation of +.30 or higher, thus explaining more than 9% or more of the observed variation).

- Clady v. County of Los Angeles (1985). This case involved an entry-level firefighter written test. The court stated, "In conclusion, the County's validation studies demonstrate legally sufficient correlation to success at the Academy and performance on the job. Courts generally accept correlation coefficients above +.30 as reliable . . . As a general principle, the greater the test's adverse impact, the higher the correlation which will be required."

- Zamlen v. City of Cleveland (1988). This case involved several different entry-level firefighter physical ability tests that had various correlation coefficients with job performance. The judge noted that, "Correlation coefficients of .30 or greater are considered high by industrial psychologists" and set a criteria of .30 to endorse the City's option of using the physical ability test as a ranking device.

Weighting Tests into Combined Scores

Tests can be weighted and combined into a composite score for each applicant. Typically, each test that is used to make the combined score is also used as a screening device (*i.e.,* with a pass/fail cutoff) before including scores from applicants into the composite score. Before using a test as a pass/fail device and as part of a weighted composite, the developer should evaluate the extent to which the KSAPCs measured by the tests are performance differentiating—especially if the weighted composite will be used for ranking applicants.

Determining a set of job-related weights to use when combining tests can be a sophisticated and socially sensitive issue. Not only are the statistical mechanics often complicated, choosing one set of weights versus another can sometimes have a very significant impact on the gender and ethnic composition of those who are hired from the final list. For these reasons, this topic should be approached with caution and practitioners should make decisions using informed judgment and input from subject-matter experts wherever possible.

There are two critical factors to consider when weighting tests into composite scores: (1) determining the weights and (2) standardizing the scores. Developing a reliability coefficient[20] for the final list of composite scores is also a critical final step if the final scores will be banded into groups of substantially equally qualified applicants. These steps are discussed below.

Generally speaking, weighting the tests that will be combined into composite scores for each applicant can be accomplished using one of three methods: *unit weighting*, weighting based on *criterion-related validity* studies, and using *content validity* weighting methods.

Unit weighting is accomplished by simply allowing each test to share an equal weight in the combined score list. Surprisingly, sometimes unit weighting produces highly effective and valid results (see the *Principles*, 2003, p. 20). This is probably because each test equally contributes to the composite score, and no test is hampered by only contributing a small part to the final score. Using unit weighting, if there are two tests, they are each weighted 50%. If there are five, each is allowed 20% weight.

If the tests that are based on one or more criterion-related validity studies are being used, the data from these studies can be used to calculate the weights for each. The steps for this method are outside the scope of this text and will not be discussed here.[21]

Using content validity methods to weight tests is probably the most common practice. Sometimes practitioners get caught up in developing computationally-intensive methods for weighting tests using job analysis data. Sometimes these procedures involve using complicated formulas that consider frequency and importance ratings for job duties and/or KSAPCs, and/or the linkages between these. While this helps some practitioners feel at ease, these methods can produce misleading results. Not only that, there are easier methods available (proposed below).

For example, consider two KSAPCs that are equally important to the job. Now assume that one is more complex than the other, so it is divided into two KSAPCs on the job analysis and the other (equally important) KSAPC remains in a single slot on the Job Analysis. When it comes time to use multiplication formulas to determine weights for the tests that are linked to these two KSAPCs, the first is likely to receive more weight *just because it was written twice on the Job Analysis*. The same problem exists if tests are mechanically linked using job duties that have this issue.

What about just providing the list of KSAPCs to a panel of Job Experts and having them distribute 100 points to indicate the relative importance of each? This method is fine, but can also present some limitations. Assume there are 20 KSAPCs and Job Experts assign importance points to each. Now assume that only 12 of these KSAPCs are actually tested by the set of tests chosen for the weighted composite. Would the weight values turn out differently if the Job Experts were allowed to review the 12 remaining KSAPCs and were asked re-assign their weighting values? Quite possibly, yes.

Another limitation with weighting tests by evaluating their relative weight from job analysis data is that sometimes different tests are linked to the same KSAPC (this can cause the weights for each test to be no longer unique and become convoluted with other tests). One final limitation is that sometimes tests are linked to a KSAPC for collecting the weight determination, but they are weak measures of the KSAPC (while others are strong, relevant linkages). For these reasons, there is a "better way" (discussed below).

Steps for Weighting Tests Using Content Validity Methods

The following steps can be taken to develop content valid weights for tests that are combined into single composite scores for each applicant:

1. Select a panel of 4 to 12 Job Experts who are truly experts in the content area and are diverse in terms of ethnicity, gender, geography, seniority (use a minimum of one year experience), and "functional areas" of the target position.

2. Provide a copy of the Job Analysis for each Job Expert. Be sure that the Job Analysis itemizes the various job duties and KSAPCs that are important or critical to the job.

3. Provide each Job Expert with a copy of each test (or a highly detailed description of the content of the test if confidentiality issues prohibit Job Experts from viewing the actual test).

4. Explain the confidential nature of the workshop, the overall goals and outcomes, and ask the Job Experts to sign confidentiality agreements.

5. Discuss and review with Job Experts the content of each test and the KSAPCs measured by each. Also discuss the extent to which certain tests may be better measures of certain KSAPCs than others. Factors such as the vulnerability of certain tests to fraud, reliability issues, and others should be discussed.

6. Provide a survey to Job Experts that asks them to distribute 100 points among the tests that will be combined. Be sure that they consider the importance levels of the KSAPCs measured by the tests, and the job duties to which they are linked, when completing this step.

7. Detect and remove outlier Job Experts from the data set.

8. Calculate the average weight for each test. These averages are the weights to use when combining the test into a composite score.

Standardizing Scores

Before individual tests can be weighted and combined, they should be *standard scored.* Standard scoring is a statistical process of *normalizing* scores and is a necessary step to place different tests on a level playing field.

Assume a developer has two tests: one with a score range of 0 – 10 and the other with a range of 0 – 50. What happens when these two tests are combined? The one with a high score range will greatly overshadow the one with the smaller range. Even if two tests have the same score range, they should still be standard scored. This is because if the tests have different means and standard deviations they will produce inaccurate results when combined unless they are first standard scored.

Standard scoring tests is a relatively simple practice. Converting raw scores into *Z scores* (a widely used form of standard scoring) can be done by simply subtracting each applicant's score from the average (mean) score of all applicants and dividing this value by the standard deviation (of all applicant total scores). After the scores for each test are standard scored, they can be multiplied by their respective weights and a final score for each applicant calculated. After this final score list has been compiled, the reliability of the new combined list can be calculated.[22]

Chapter 5 – Steps for Developing a Content-Valid, Job Knowledge Written Test

Job knowledge can be defined as "...the accumulation of facts, principles, concepts and other pieces of information that are considered important in the performance of one's job" (Dye, Reck, & McDaniel, 1993, p. 153). As applied to written tests in the personnel setting, knowledge can be categorized as: *declarative knowledge* – knowledge of technical information; or *procedural knowledge* – knowledge of the processes and judgmental criteria required to perform correctly and efficiently on the job (Hunter, 1984; Dye, Reck, & McDaniel, 1993).

While job knowledge is not typically critical for many entry-level positions, it clearly has its place in many supervisory positions where having a command of certain knowledge areas is essential for job performance. For example, if a Fire Captain, responsible for instructing firefighters who have been deployed to extinguish a house fire, does not possess a mastery-level of knowledge required for the task, the safety of the firefighters and the public could be in jeopardy. It is not feasible to require a Fire Captain in this position to refer to textbooks and determine the best course of action, but rather that he or she must have the particular knowledge memorized.

As described previously, there are a variety of steps that should be followed to ensure that a job knowledge written test is developed and utilized properly. Depending upon the size and type of the employer, they may be faced with litigation from state or federal regulatory agencies or a private plaintiff attorney. Each year, employers accused of utilizing tests that have adverse impact spend millions of dollars defending litigated promotional processes.[23]

In litigation settings, addressing these standards is typically conducted by completing a validation study (using any of the acceptable types of validity). This section outlines seven steps for developing a job-related and court-defensible process for creating a *content*-valid, job knowledge written test used for hiring or promoting employees.

The seven steps below are designed to address the essential requirements based on the *Uniform Guidelines* (1978), the *Principles* (2003), and the *Standards* (1999)[24]:

1. Conduct a job analysis;
2. Develop a selection plan;
3. Identify test plan goals;
4. Develop the test content;
5. Validate the test;
6. Compile the test; and,
7. Conduct post-administration analyses.

Step 1: Conduct a Job Analysis

The foundational requirement for developing a content-valid, job knowledge written test is a current and thorough job analysis for the target position. Brief 1-2 page "job descriptions" are almost never sufficient for showing validation under the *Uniform Guidelines* unless, at a bare minimum, they include:

- Job Expert input and/or review;
- Job duties and KSAPCs that are essential for the job;
- Operationally defined KSAPCs.

In practice, where validity is required, updated job analyses typically need to be developed. Ideally, creating a *Uniform Guidelines*-style job analysis would include the following ratings for **job duties**.

Frequency (*Uniform Guidelines*, Section 15B3; 14D4)[25]

This duty is performed *(Select one option from below)* by me or other active (target position) in my department.

1. **annually** or less often
2. **semi-annually** (approx. 2 times/year)
3. **quarterly** (approx. 4 times/year)
4. **monthly** (approx. 1 time/month)
5. **bi-weekly** (approx. every 2 weeks)
6. **weekly** (approx. 1 time/week)
7. **semi-weekly** (approx. 2 times/week)
8. **daily/infrequently** (approx. 1 to 6 times/day)
9. **daily/frequently** (approx. 7 or more times/day)

Importance (*Uniform Guidelines*, Section 14C1, 2, 4; 14D2, 3; 15C3, 4, 5; 15D3)

Competent performance of this duty is *(Select one option from below)* for the job of (target position) in my department.

1. **not important: minor** significance to the performance of the job.
2. **of some importance: somewhat useful and/or** meaningful to the performance of the job. Improper performance may result in **slight** negative consequences.
3. **important**: **useful and/or meaningful** to the performance of the job. Improper performance may result in **moderate** negative consequences.

4. **critical: necessary** for the performance of the job. Improper performance may result in **serious** negative consequences.
5. **very critical: necessary** for the performance of the job, and with more extreme consequences. Improper performance may result in **very serious** negative consequences.

Ideally, creating a *Uniform Guidelines*-style job analysis requires that each **KSAPC** has the following ratings:

Frequency (*Uniform Guidelines*, Section 15B3; 14D4)[16]

This KSAPC is performed *(Select one option from below)* by me or other active (target position) in my department.

1. **annually** or less often
2. **semi-annually** (approx. 2 times/year)
3. **quarterly** (approx. 4 times/year)
4. **monthly** (approx. 1 time/month)
5. **bi-weekly** (approx. every 2 weeks)
6. **weekly** (approx. 1 time/week)
7. **semi-weekly** (approx. 2 times/week)
8. **daily/infrequently** (approx. 1 to 6 times/day)
9. **daily/frequently** (approx. 7 or more times/day)

Importance (*Uniform Guidelines*, Section 14C1, 2, 4; 14D2, 3; 15C3, 4, 5; 15D3)

This KSAPC is *(Select one option from below)* for the job of (target position) in my department.

1. **not important: minor** significance to the performance of the job.
2. **of some importance: somewhat useful and/or** meaningful to the performance of the job. Not

possessing adequate levels of this KSAPC may result in **slight** negative consequences.
3. **important: useful and/or meaningful** to the performance of the job. Not possessing adequate levels of this KSAPC may result in **moderate** negative consequences.
4. **critical: necessary** for the performance of the job. Not possessing adequate levels of this KSAPC may result in **serious** negative consequences.
5. **very critical: necessary** for the performance of the job, and with more extreme consequences. Not possessing adequate levels of this KSAPC may result in **very serious** negative consequences.

Differentiating "Best Worker" Ratings (*Uniform Guidelines*, Section 14C9)

Possessing **above-minimum levels** of this KSAPC makes *(Select one option from below)* difference in overall job performance.

1. no
2. little
3. some
4. a significant
5. a very significant

Note: Obtaining ratings on the "Best Worker" scale is not necessary if the job knowledge written test will be used only on a pass/fail basis (rather than ranking final test results).

When Needed (*Uniform Guidelines*, Section 5F; 14C1)

Possessing *(Select one option from below)* of this KSAPC is needed upon entry to the job for the (target position) position in your department.

1. none or very little

2. some (less than half)
3. most (more than half)
4. all or almost all

In addition to these four KSAPC rating scales, it is recommended that a **mastery level** scale be used when validating written job knowledge tests. The data from these ratings are useful for choosing the job knowledges that should be included in a written job knowledge test, and are useful for addressing Section 14C4 of the *Uniform Guidelines*, which require that job knowledges measured on a test be ". . . operationally defined as that body of learned information which is used in and is a necessary prerequisite for observable aspects of work behavior of the job." It is recommended to use an average rating threshold of 3.0 on the mastery-level scale for selecting which job knowledges to include on job knowledge tests. A sample mastery level scale is listed below:

Mastery Level (*Uniform Guidelines*, Section 14C4)

A *(Select one option from below)* level of this job knowledge is necessary for successful job performance.

1. **low**: none or only a few general concepts or specifics available in memory in none or only a few circumstances without referencing materials or asking questions.
2. **familiarity**: have some general concepts and some specifics available in memory in some circumstances without referencing materials or asking questions.
3. **working knowledge**: have most general concepts and most specifics available in memory in most circumstances without referencing materials or asking questions.

4. **mastery**: have almost all general concepts and almost all specifics available in memory in almost all circumstances without referencing materials or asking questions.

Finally, a duty/KSAPC **linkage scale** should be used to ensure that the KSAPCs are necessary to the performance of important job duties. A sample duty/KSAPC linkage scale is provided below:

Duty/KSAPC Linkages (*Uniform Guidelines*, Section 14C4)

This KSAPC is ________ to the performance of this duty.

1. not important
2. of minor importance
3. important
4. of major importance
5. critically important

When Job Experts identify KSAPCs necessary for the job, it is helpful if they are written in a way that maximizes the likelihood of job duty linkages. When KSAPCs fail to provide enough content to link to job duties, their inclusion in a job analysis is limited. Listed below are examples of a poorly written and a well written KSAPC from a firefighter job analysis:

- Example of a **poorly-written** KSAPC: *Knowledge of ventilation practices.*

- Example of a **well-written** KSAPC: *Knowledge of ventilation practices and techniques to release contained heat, smoke, and gases in order to enter a building. Includes application of appropriate fire* *suppression techniques and equipment (including manual and power tools and ventilation fans).*

Step 2: Develop a Selection Plan

The first step in developing a selection plan is to review the KSAPCs from the job analysis and design a plan for measuring the essential KSAPCs using various selection procedures (particularly, knowledge areas). At a minimum, the knowledge areas selected for the test should be important, necessary on the first day of the job, required at some level of mastery (rather than easily looked up without hindrance on the job), and appropriately measured using a written test format. Job knowledges that meet these criteria are selected for inclusion in the "Test Plan" below.

Step 3: Identify Test Plan Goals

Once the KSAPCs that will be measured on the test have been identified, the test sources relevant for the knowledges should be identified. Review relevant job-related materials and discuss the target job in considerable detail with Job Experts. This will focus attention on job- specific information for the job under analysis. Review the knowledges that meet the necessary criteria and determine which sources and/or textbooks are best suited to measure the various knowledges. It is imperative that the selected sources do not contradict one another in content.

Once the test sources have been identified, determine whether or not preparatory materials will be offered to the applicants. If preparatory materials are used, ensure that the materials are current, specific, and released to all applicants taking the test. In addition to preparatory materials, determine if preparatory sessions will be offered to the applicants.

Use of preparatory sessions appears to be beneficial to both minority and non-minority applicants, although they do not consistently reduce adverse impact (Sackett, Schmitt, Ellingson, & Kabin, 2001). If study sessions are conducted, make every attempt to schedule study sessions at a location that is geographically convenient to all applicants and that they are offered at a reasonable time of day. Invite all applicants to attend and provide plenty of notice of the dates and times.

Following the identification of the knowledge areas and source materials that will be used to develop the job knowledge written test, identify the number of test items that will be included on the test. Be sure to include enough items to ensure high test reliability. Typically, job knowledge tests that are made up of similar job knowledge domains will generate reliability levels in the high .80s to the low .90s when they include 80 items or more.

Consider using Job Expert input to determine *internal weights* for the written test. Provide Job Experts with the list of knowledges to be measured and ask experts to distribute 100 points among the knowledges to obtain a balanced written test. See Table 5 for a sample of a knowledge weighting survey used to develop a written test for certifying firefighters (this type of test would be used by fire departments that hire only pre-trained firefighters into entry-level positions).

Table 5. Firefighter Certification Test Development Survey

<table>
<tr><th colspan="2">Firefighter Certification Test Development Survey</th></tr>
<tr><td colspan="2">Job Expert Name: ______________________ Date:______ ____________</td></tr>
<tr><td colspan="2">Instructions: Assume that you have $100 to “buy” the perfect firefighter for your department (based only on job knowledge qualifications—assume other important areas such as physical abilities and interpersonal skills have already been tested). How much money would you spend in the following job knowledge sources areas to “buy the most qualified firefighter” for your department? Be sure that your allocations equal exactly $100.</td></tr>
<tr><td>Knowledge Sources</td><td>Dollars You Would Spend to “Buy” the Perfect Firefighter</td></tr>
<tr><td>Pumping Apparatus Driver/Operator</td><td></td></tr>
<tr><td>Principles of Vehicle Extrication</td><td></td></tr>
<tr><td>Fire Department Company Officer</td><td></td></tr>
<tr><td>Fire and Emergency Services Instructor</td><td></td></tr>
<tr><td>Aerial Apparatus Driver/Operator</td><td></td></tr>
<tr><td>Essentials of Firefighting</td><td></td></tr>
<tr><td>Rapid Intervention Teams</td><td></td></tr>
<tr><td>The Source Book for Fire Company Training Evolutions</td><td></td></tr>
<tr><td>Fire Inspection and Code Enforcement</td><td></td></tr>
<tr><td>Hazardous Materials</td><td></td></tr>
<tr><td>TOTAL (must equal $100)</td><td></td></tr>
</table>

Attempt to obtain *adequate sampling* of the various knowledges and ensure that there are a sufficient number of items developed to effectively measure each knowledge at the desired level. Note that some knowledges will require more items than others for making a "sufficiently deep" assessment. The test should be internally weighted in a way that ensures a sufficient measurement of the relevant knowledge areas.

Following the determination of the length of the test and the number of items to be derived from each source, determine the *types* of items that will be included on the test. One helpful tool is a process-by-content matrix to ensure adequate sampling of job knowledge content areas and problem-solving processes. Problem-solving levels include:

- *Knowledge* of terminology
- *Understanding* of principles
- *Application* of knowledge to new situations

While knowledge of terminology is important, the *understanding* and *application* of principles may be considered of primary importance. The job knowledge written test should include test items that ensure the applicants can define important terms related to the job *and* be able to apply their knowledge to answer more complex questions. Job Experts should consider how the knowledge is applied and required on the job (*e.g.,* at the "definition" level only, the "principle" level where the basic concepts of the areas assessed need to be understood, and the "application" level where a greater level of mastery is required) when determining the types of items to be included on the final test form (see Table 6 for a sample process-by-content matrix for a Fire Captain written test).

Table 6. Process-by-Content Matrix: Fire Captain

Process-by-Content Matrix: Fire Captain				
Source	**Definition**	**Principle**	**Application**	**Total**
1. Brady Emergency Care	4	10	20	34
2. Brunacini, Fire Command	3	7	13	23
3. Fire Chief's Handbook	3	10	17	30
4. IFSTA Building Construction	1	3	6	10
5. IFSTA Wildland Firefighting	4	5	9	18
6. IFSTA Fire Service Rescue	4	6	10	20
7. IFSTA Essentials	2	2	6	10
8. IFSTA Chief Officer	0	1	1	2
9. IFSTA Hazardous Materials	0	1	2	3
Total	21	45	84	150

Step 4: Develop the Test Content

After the number and types of test items to be developed have been determined, select a diverse panel of 4 to 10 Job Experts (who have a minimum of one year experience) to review the test plan to ensure compliance with the parameters. Have each Job Expert sign a "Confidentiality Form." If the Job Experts are going to write the test items, provide item-writing training and have Job Experts write, exchange, and review the items.

Once the Job Experts have written the items to be included in the test bank, ensure proper grammar, style, and consistency. Additionally, make certain that the test plan requirements are met. Once the bank of test items has been created, provide the final test version to the panel of Job Experts for the validation process (the next step).

Step 5: Validate the Test

The next step (validating the test) is the most important one in the entire process. The validation process is completed by convening a panel of qualified Job Experts and having them provide ratings on several factors relevant to the test items and the Job Analysis for the target position. While the questions on the validation survey will vary widely across departments and testing consultants, the basic goal is the same: to establish a connection between the test items and the target job.

Whatever system is used, it is critical to include survey questions that will address the fundamental requirements of the federal *Uniform Guidelines* and professional testing guidelines (e.g., the *Principles* and the *Standards*). Figure 1 below provides a survey that is used as part of the *Test Validation & Analysis Program* (TVAP®) published by Biddle Consulting Group, Inc. (BCG, 2011). This survey includes the key factors that should be evaluated by Job Experts during a written test validation process.

Figure 1. Test Validation Survey (from the BCG Test Validation & Analysis Program)

Name: ______________________ Test: ____________ Date: ______________

		Test Item Number		
		#1	#2	#3...
1. Does the item STEM:				
A. READ WELL? (Is it CLEAR and UNDERSTANDABLE?)	N/Y	N Y	N Y	N Y
B. Provide SUFFICIENT INFORMATION to answer correctly?	N/Y	N Y	N Y	N Y
2. Are the DISTRACTORS:				
A. Similar in difficulty?	N/Y	N Y	N Y	N Y
B. Distinct?	N/Y	N Y	N Y	N Y
C. Incorrect, yet plausible?	N/Y	N Y	N Y	N Y
D. Similar in length?	N/Y	N Y	N Y	N Y
E. Correctly matching to the stem?	N/Y	N Y	N Y	N Y
3. Is the key CORRECT IN ALL CIRCUMSTANCES?	N/Y	N Y	N Y	N Y
4. Is this item FREE OF PROVIDING CLUES to other items?	N/Y	N Y	N Y	N Y
5. Is this item FREE FROM UNNECESSARY COMPLEXITIES?	N/Y	N Y	N Y	N Y
6. What percent of MINIMALLY QUALIFIED APPLICANTS would you expect to answer this item correctly?	%			
7. Is this item FAIR to all groups of people?	N/Y	N Y	N Y	N Y
8. What JOB DUTY is represented by this item?	# from Job Analysis			
9. What KSAPC is being measured by this item?				
10. Does the item measure a part of the KSAPC that is NECESSARY ON THE FIRST DAY of the job?	N/Y	N Y	N Y	N Y
Only Answer Items 11-14 Below for Job Knowledge Test Items				
11. Is the item BASED ON CURRENT INFORMATION?	N/Y	N Y	N Y	N Y
12. How important is it that the knowledge tested be MEMORIZED? 0 - NOT NECESSARY: can be looked up without impacting job performance 1 - IMPORTANT: negative job impact is LIKELY if it had to be looked up 2 - ESSENTIAL: negative job impact is MOST LIKELY if it had to be looked up	0-2			
13. Does the LEVEL OF DIFFICULTY of the item correspond to the level of difficulty of the knowledge as used on the job?	N/Y	N Y	N Y	N Y
14. How serious are the CONSEQUENCES if the applicant does not possess the knowledge required to answer this item correctly? 0 - LITTLE or NO consequences 1 - MODERATE consequences 2 - SEVERE consequences	0-2			

After the ratings are collected from Job Experts for each item on the test using the TVAP survey, the program applies criteria derived from court cases, the federal *Uniform Guidelines*, and professional testing standards to mark each item as "valid" or "not valid" and sets a passing score for the test using the modified Angoff procedure (see below). Whatever system is used, the fundamental questions displayed in Figure 1 should be included and addressed by the validation and rating process.

Additionally, have the Job Experts identify an appropriate time limit for the test. A common rule-of-thumb used by practitioners to determine a written test cutoff time is to allow one minute per test item plus thirty additional minutes (*e.g.*, a 150-item test would yield a three hour time limit).[26] A reasonable time limit would allow for at least 95% of the applicants to complete the test within the time limit.[27]

Step 6: Compile the Test

The test items that survive the validation review process should be compiled into a final test form and an *unmodified Angoff* (the test cutoff that will be slightly downward adjusted after the administration based on the reliability of the test) should be set by averaging the "Angoff" ratings given by Job Experts. Raters whose ratings are statistically different from other raters should be identified (by evaluating rater reliability and high/low rater bias) and removed.

Step 7: Post-Administration Analyses

Following the administration of the job knowledge written test, conduct an item-level analysis of the test results to evaluate the item-level qualities (such as the point-biserial, difficulty level, and Differential Item Functioning [DIF] of each item).

When evaluating whether or no to discard an item due to DIF, consider the following excerpt from Hearn v. City of

Jackson (Aug. 7, 2003)[25] where DIF was being considered for a job knowledge test:

> Plaintiffs suggest in their post-trial memorandum that the test is subject to challenge on the basis that they failed to perform a DIF analysis to determine whether, and if so on which items, blacks performed more poorly than whites, so that an effort could have been made to reduce adverse impact by eliminating those items on which blacks performed more poorly. . . Dr. Landy testified that the consensus of professional opinion is that DIF modification of tests is not a good idea because it reduces the validity of the examination. . . Dr. Landy explained: The problem with [DIF] is, suppose one of those items is a knowledge item and has to do with an issue like Miranda or an issue in the preservation of evidence or a hostage situation. You are going to take that item out only because whites answer it more correctly than blacks do, in spite of the fact that you would really want a sergeant to know this [issue] because the sergeant is going to supervise. A police officer is going to count on that officer to tell him or her what to do. So you are reducing the validity of the exam just for the sake of making sure that there are no items in which whites and blacks do differentially, or DIF, and he is assuming that the reason that 65 percent of the blacks got it right and 70 percent of the whites got it right was that it is an unfair item rather than, hey, maybe two or three whites or two or three blacks studied more or less that section of general orders.

Certainly this excerpt provides some good arguments against discarding items based only on DIF analyses *for job knowledge tests*. Tests measuring skills and abilities, however, may be more prone to DIF issues. These issues should be carefully considered before removing items from a test.

After conducting the item-level analysis and removing items that do not comply with acceptable ranges, conduct a test-level analysis to assess descriptive and psychometric statistics (*e.g.*, reliability, standard deviation, etc.). After removing any test items with poor psychometric properties (due to miskeying, etc.), determine the final cutoff for the test by subtracting one, two, or three Standard Errors of Measurement (SEMs) or Conditional Standard Errors of Measurement (C-SEMs) from the unmodified Angoff level.[28]

In summary, developing a content-valid, job knowledge written test for hiring/promoting employees (where the job requires testing for critical job knowledge areas) is the safest route to avoid potential litigation. If the test has adverse impact – validate. Pay particular attention in addressing the *Uniform Guidelines*, *Principles*, and *Standards* (in that order, based on the weight they are typically given in court), and remember that a house is only as strong as its foundation. Be sure to base everything on a solid job analysis.

Chapter 6 – Developing and Validating "Work Sample" Physical Ability Tests

Let's face it: the firefighter occupation is physically demanding. Fire departments often struggle with installing defensible testing instruments that can select candidates who will succeed in the physical aspects of the firefighter job.

For example, some departments install strict cardiovascular fitness tests that use stair climbing machines or stationary bicycles to estimate the applicant's maximum V02 threshold.[29] Such tests may lead to disgruntled applicants who do not perceive a direct connection between the test and the requirements of the job (*i.e.,* a lack of "face validity," which can lead to low perceptions of fairness). In addition, a test that measures an applicant's "physiological or biological responses to performance" is classified by the EEOC (under the Americans with Disabilities Act (ADA) of 1990) as a *medical examination*, which means that it can only be administered after a contingent job offer has been made.[30] The scoring and use of this type of test becomes even more challenging because they typically require using gender and age in the scoring formulae (Siconolfi, Garber, Lasater, & Carleton, 1985) which is a specific violation of the 1991 Civil Rights Act when it comes to employment testing.[31]

Other departments install only *static strength tests* that measure whether an applicant is capable of lifting or manipulating the weights that are routinely handled on the job, and leave stamina unmeasured in the hiring process. Still other departments do not install any form of testing, and leave it to chance as to whether candidates will be able to perform rigorous job requirements. Neither of these solutions alone will likely serve the department's best interests when hiring for the physically-demanding firefighter job.

The best way to ensure that candidates are job-ready is to develop a *work sample test* that replicates and mirrors a "vital snapshot" of the firefighter job. In this way, whatever combination of strength and fitness that is required for the actual job is mirrored on the pre-employment test (as much of each that can feasibly be included on a pre-employment test). Research strongly suggests there will be fewer dissatisfied test takers if the content of a test is transparently similar to the content of the job, such that those who fail the test would realize that they would not have successfully performed the job. This is because work sample tests typically have a higher *perception of applicant fairness*. Because they look and feel more like the actual job, applicants are less likely to walk away disgruntled from their testing experience—even if they fail the test.

With that in mind, it is frequently helpful if the content and context of the physical ability and/or work sample test events mimic critical or important work behaviors that constitute most of the job. In the words of Section 14C4 of the federal *Uniform Guidelines*, "the closer the content and context of the selection procedure are to work samples or work behaviors, the stronger is the basis for showing content validity." The following is a description of a content-related validity strategy for validating a physical ability test.

Steps for Developing a Physical Ability Test Using Content Validity

1. Conduct a thorough job analysis that focuses on the physical aspects of the job.

2. Identify the parts of the job (*i.e.*, job duties, or sets of job duties) that are typically performed in rapid succession that collectively require *continuous physical exertion* for over 10 minutes. This can include one job duty repeatedly performed where a rapid work pace is required on the job (*e.g.*, loading or unloading hose), or a set of unrelated job duties where a rapid pace is required for physically-demanding job duties (*e.g.*, pulling hoses, then raising the fly section of a ladder).

3. Work with supervisors and trainers to assemble a continuously-timed, multiple-event job simulation physical ability test (PAT). The PAT events included need to be those where a rapid (but safe) working pace is important. If you wish to use physical ability testing for parts of the job that are not typically performed in rapid succession, then work with supervisors and trainers to assemble discrete test events for measuring the ability to perform those parts of the job.

4. Run a representative sample of Job Experts (*e.g.*, 20-30) through the PAT and administer a validation survey that collects the following information from each:

    a. Actual PAT completion time. If you are using a continuously-timed, multiple-event job simulation PAT, then collect the time it takes each Job Expert to complete all of the events combined. If you are using a single-event PAT, or a series of single-event PATs, then collect the time it takes each Job Expert to complete each test event.

b. Opinion time for a minimally-qualified applicant to complete the PAT (*e.g.,* Job Experts could be asked, "Given your time to complete the PAT, your current fitness level, and your level of job experience, what time should a minimally-qualified applicant score when taking this PAT?").

c. Their Yes/No answer to the following questions:
    i. Does the PAT measure skills/abilities that are important/critical (essential for the performance of the job)?
    ii. Does the PAT measure skills/abilities that are necessary on the first day of the job (*i.e.,* before training)?
    iii. Does the PAT replicate/simulate actual work behaviors in a manner, setting, and level of complexity similar to the job?
    iv. Do the events in the PAT need to be completed on the job in a <u>rapid and safe</u> manner (*i.e.,* is speed important)?
    v. Are the weights and distances involved in the PAT representative of the job?
    vi. Is the duration that the objects/equipment are typically carried or handled in the PAT similar to what is required of a single person on the job?
    vii. Is the PAT free from any "special techniques" that are learned on the job that allow current job incumbents to perform the PAT events better than an applicant could (that are not demonstrated to the applicants prior to taking the PAT)?

viii. Does the PAT require the same or less exertion of the applicant than is required on the job?

5. Analyze the Job Expert data gathered from Step 4:

   a. First, analyze the Yes/No ratings gathered from Step 4c (above). At least 70% of the Job Experts must answer "Yes" to each question. If this is not the case, go back to the drawing board and re-design the test as necessary. Then re-survey the Job Experts until at least 70% of the Job Experts endorse each.
   b. Next, compute the average of the opinion times gathered in Step 4b. Use the SEM computed in Step 6 below to adjust their average opinion time (by adding 1 or 2 SEMs) to set the final cutoff score (*i.e.,* the amount of time that is allowed to successfully complete the test event).
   c. Compute the average Job Expert PAT time (from 4a above) and add 1.645[32] Standard Errors of Difference (SEDs) to this average.[33] This score level may help identify[34] the lowest score boundary that constitutes the "normal expectations of acceptable proficiency in the workforce" specified by the *Uniform Guidelines* (see Section 5H). Ideally, the final cutoff score (set using the adjusted opinion times above) should be close to this score level.

6. Conduct a test-retest study to determine the Standard Error of Measurement (SEM) of the PAT. This requires having 60+ applicants or incumbents taking the PAT twice (separated by 1-2 weeks), and correlating the two scores to obtain the test-retest reliability estimate (r_{tt}). This value is used along with the standard deviation of the sample to compute the SEM using the formula: *SEM*

$= \sigma_x(1 - r_{tt})^{1/2}$ where σ_x is the standard deviation of test scores and r_{tt} is the test-retest reliability. For example, if the test-retest reliability is .84 and the standard deviation is 10, the SEM would be 4.0 (in Excel: = 10*(sqrt(1-.84)). As an example, the SEM for the FPSI Work Sample PAT is r = .8015.

Any pace, distance, weights, or other limitations used during testing must be job-related (*i.e.,* related to actual pace, distances, weights, or other limitations that are required on the job). For example, if an employee on the job would have up to three minutes to move something from point A to point B, then the job candidate should not be required to perform this same task in less time during the test.[35] Similarly, if an employee is expected to carry an object a certain distance on the job, but is allowed to briefly place that object on the ground to rest and/or change their grip, the test taker should also be allowed to similarly rest and/or change their grip if required to carry an item during a test. In other words, the test should be similar in difficulty and execution to the actual job, and should not require the test taker to carry more weight, move something a longer distance, or perform work that is substantially more difficult than on the actual job.

Test events that mimic work behaviors do not require a statistical examination of test performance and job performance to be conducted. Alternatively, test events that do not mimic work behaviors, but which are predictive of job performance, can be used if a statistically significant relationship can be shown between performance on the test and performance on the job. To use this type of testing, a relatively large number of job candidates or current employees (at least 150 is recommended, though meaningful studies can be conducted with smaller samples if the relationship between the test and job performance is relatively strong) must take the test and their test scores be statistically compared to their work performance. This process utilizes a *criterion-related validation strategy*, and is described below.

Steps for Developing a Physical Ability Test Using Criterion-related Validity

1. Complete Steps 1-3 outlined in the content validity section above.

2. If a *concurrent* criterion-related validation strategy (where the PAT will be administered to current job incumbents from the target position) will be used, administer the PAT to 150 current incumbents. The benefit of conducting a concurrent study is that it is fast – the researcher will find out quickly if the PAT is valid (*i.e.,* significantly correlated to job performance). The drawback is that concurrent studies sometimes have less "power" than the predictive studies (because of range restriction in the post-screened worker population involved in the study). This might result in a situation where the study does not reveal a significant correlation that actually exists in the greater population and would have shown up if a larger study had been conducted. Another drawback of a concurrent study is that it requires off-the-job time from incumbents to complete the PAT.

3. If a *predictive* criterion-related validation strategy (where the PAT will be administered to onboarding applicants and subsequently rated on job performance) will be used, administer the PAT to 150 applicants who are subsequently hired (*e.g.,* 500 tested, 150 hired). One of the benefits of using a predictive study is that it may

have higher power (increased likelihood of finding significance) because the employer is testing a broader ability range of applicants (compared to the post-tested employee group used in a concurrent study). Another advantage is that a predictive study is "passive" – it does not take existing employees off the job. The drawback is that it takes time to test and finally hire 150+ applicants who are subsequently rated on job performance.

4. Develop a Job Performance Rating Survey (JPRS) to be used for gathering job performance ratings (*e.g.*, using a 1-10 rating scale) from supervisors (and/or co-workers) for each incumbent who completed the PAT. Several aspects of job performance (*e.g.*, 5-10) that are suspected of being related to the PAT scores should be included. A word of caution on this part of the study: Supervisors or co-workers are often hesitant to provide any of their team members with "average" or "below average," and this tendency can easily impede the study because correlations are strongest when the "low" "middle" and "high" performance ranges are represented (on both the test and the job performance sides of the study). For this reason, it is strongly advised to inform study participants that the performance ratings will only be used for test validation purposes.

5. Administer the JPRS and collect the ratings.

6. Complete the statistical analyses and determine whether the PAT is significantly correlated to one or more job performance dimensions.

7. Set the cutoff at a level that represents the performance levels required for the job, being alert to any corresponding levels of adverse impact.

When using either a content- or criterion-related validation strategy, physical ability and work sample test events should be designed to be as simple as practical (unless the job analysis shows that a minimally-qualified employee on the first day of the job prior to training would need to be able to learn complex work-related tasks in a short period of time and then be able to perform the tasks just learned). In other words, each event should include as few steps and/or procedures for the job candidate to follow as possible, unless more complexity is justified by the job analysis. Simplicity of the test events helps minimize confusion and aides with scoring. If possible, longer or more complex events should be broken into shorter, separate parts. Also, the amount of time and information that a candidate is provided to learn how to perform the event (and is allowed to practice and ask questions about that event) should increase as the complexity of the event increases.

Administering the Test

Test administrators should be trained about the testing process prior to their being permitted to administer the test. Furthermore, it is recommended that, if testing will take place over any period of time or uses multiple administrators, a lead administrator be appointed to oversee the test's administration to ensure continuity between testing sessions. In addition, test administrators should faithfully follow the test event description plan that has been developed for the administration and scoring of each event. Deviating from the plan can result in increased potential liability to the employer.

Prior to the administration of each test event, candidates should be informed of the contents of that event. To help ensure standardization, which is fundamental for fair and valid testing, it is best if the administrator reads from a script word-for-word, or plays a recording, that describes each event before job candidates attempt those events. The administrator should also demonstrate (or show a video demonstrating) each event (including the different techniques that may be used for successfully performing the required task during each event).

In general, if a candidate appears confused or frustrated when taking the test, the administrator should ask, "Do you need me to repeat the instructions?" If the candidate says "Yes," or if a candidate directly asks for additional instructions, the administrator should provide appropriate information. An exception to this would be if the ability to follow instructions is part of the testing criteria of the test being administered. If that is the case, have a plan in place for how to address this issue in a job-related fashion (based on the job analysis and input from the job experts) in advance of testing.

Since many of the job candidates will be performing the events required during testing for the first time, it would make sense (in most situations) that they be given a longer amount of time, or allowed greater flexibility, to perform the task than if they were being performed by someone who has been performing the job for a relatively long period of time. Similarly, the number of attempts to complete an event should be a realistic number that could be used to determine whether that job candidate could successfully perform a similar task on the job. In the interest of fairness, test takers should generally be allowed more than one attempt to complete an event, unless it is obvious that injury or harm would occur to the test taker (or others), and/or expensive or non-replaceable company property would be seriously damaged if another attempt was allowed. If testing is stopped because of the likelihood that injury or harm would occur if another attempt was allowed, this should be documented in detail.

In some instances, it might be acceptable to deduct points if instructions need to be provided to the test taker more than once during this type of test. However, deducting points or other penalties for this type of activity must be job-related and justified in relationship to how the job is actually performed. For example, if there is no penalty on the job if an employee asks questions or clarifications when learning to perform a task on the job, there should not be any penalty for this behavior during the test. Conversely, if the job requires that an employee learn a task and perform that task on the job without additional instructions, then taking this into consideration during testing may be justified.

When administering the physical ability and/or work-sample test to candidates, the paramount concern should be the safety of each candidate. Safety can be promoted by ensuring that each candidate is shown the proper way(s) to perform each event prior to their taking the test, and by carefully observing the test takers during the events when appropriate.

Also, if relevant, the administrator should stress the importance of safety to the test taker before the test is administered. In addition, it might be helpful to provide a demonstration of how the test taker could safely use, maneuver, lift, carry, and/or move the materials during testing. To minimize potential injuries or problems during testing, it is strongly recommended that the test taker be allowed a reasonable period of time to practice lifting and/or carrying the materials/devices to be handled during each event. Administrators should also explain to the candidate that, since safety is a primary concern, they will be disqualified if they do not follow the safety rules and/or safe working practices that have been explained to them. It might be helpful to provide a printed copy of any safety rules or safe work practices that will be used as disqualifications during the test to the job candidates in advance of the test. However, even if a printed copy of the safety rules and safe work practices are given to the candidates in advance, those rules should again be read, explained, and/or demonstrated to the candidates at the time of testing.[36] Employers should consider that any actions they take during recruitment and testing sends a message to potential employees about the culture of the organization.

Testing should be immediately stopped if a candidate fails to follow a safety rule and/or safe working practice. Explain to that candidate again how the task should be safely performed and, if appropriate, provide another demonstration. Allow the candidate to continue testing unless it is obvious that injury or harm would occur to the test taker (or others), and/or company property would be seriously damaged, in which case another attempt should not be allowed. Again, if the testing process is stopped because of the likelihood that injury or serious damage would occur if the test was continued, this should be documented in detail.

If a candidate violates a safety rule or if, during the physical ability or work sample task event, they perform the task in a way that the administrator feels demonstrates that they do not possess the level of safety-related knowledge that a minimally-qualified, entry-level employee should possess, a complete and accurate description of that violation or unsafe work behavior should be presented to a panel of target-job experts (and/or supervisors and/or trainers of the target job) or safety-committee members for evaluation.[37] Those experts shall determine whether the violation or unsafe work practice indicated that the candidate does not possess the level of safety-related knowledge or ability that a minimally-qualified, entry-level employee should possess. The job candidate will be disqualified if it is determined by the panel that their performance indicated they do not possess the level of safety-related knowledge or ability that a minimally-qualified, entry-level employee for the target job should possess prior to any training or on-the-job experience with the employer who is doing the testing.

Because the test will likely be physically strenuous, it is recommended that candidates be required to sign a waiver of liability (which must be completed and signed before a candidate is allowed to take the test). The liability release form should include a description of the test so that the candidate can make an intelligent waiver. If the employer so decides, candidates may also be required to obtain a medical release prior to testing. First aid and/or medical assistance should also be available at the testing site and/or readily available.

If a candidate appears to be injured during testing, stop the test and ask them, “Are you injured?” If the candidate says “Yes” and/or it is obvious to the test administrator that they have injured themselves, stop the event and obtain assistance immediately. Develop a plan in advance as to how to respond to injuries that may occur during testing and make certain all test administrators are aware of the emergency response plan if an injury should occur.

The starting and ending points for each event, and/or the path that the test taker should take during testing, should be clearly marked for the test taker to see. For example, if the candidate can be penalized for traveling outside of a certain path during a test event, (1) the path limits should be clearly marked so the test taker knows their limits when performing a test event, and (2) the limits should be based upon solid, job-related reasons (for instance, the narrow path that must be followed when the job is actually performed which travels between two pieces of closely-aligned equipment; or that an employee or another person might be injured on the job if a certain path is not carefully followed). It is helpful to photograph the test course to document that these steps have been taken.

In general, spectators during testing should not be allowed. However, even if there are no spectators, the candidates themselves will be observing each other perform the events. All observers and candidates should be instructed not to cheer, jeer, whistle, yell, signal, or in any way interfere with a fellow candidate's performance of an event. That being said, the test administrator is encouraged to provide a moderate, consistent level of encouragement, support, safety reminders, and/or instructions to all candidates.

If the testing requires strenuous physical activity, it is recommended that candidates be asked to remain at the testing site until they have sufficiently recovered from the testing process. This could potentially reduce claims related to testing. In general, candidates should not be permitted to leave the testing process until:

1. They have correctly performed the test events (administrators should make notes of any observable deficiencies).

2. The candidate says they wish to stop the test or cannot complete the event; the candidate is then disqualified.

3. The candidate has attempted, but did not successfully complete the event or test; the candidate is then disqualified.

Candidates who wish to leave prior to any of these three conditions should be asked, if possible, to sign a document indicating they have voluntarily withdrawn from the selection process. The administrator should carefully document the circumstances if the candidate refuses to sign such a document before leaving.

Scoring Physical Ability Tests

Scoring accuracy and fairness to all candidates can be promoted by implementing a standardized approach for the administration and scoring of each event (*i.e.,* sending all candidates through the same events, in the same sequence, and with the same instructions) and, if possible, utilizing multiple scorers (*e.g.,* two individuals with stop watches). Also, clear and unambiguous, observable criteria must be used when determining whether someone passes or fails the test event in the same way for scoring each and every test taker. Ambiguous criteria, such as whether the test taker "appeared to be struggling," "was breathing hard," or "had to stand on the tips of their toes when performing the task," are not acceptable for scoring purposes.

The final decision must be made as to whether the test taker successfully completed the task required or not, within the defined observable criteria (such as being able to carry an object which weighs the same or less than the weight carried on the job; a job-related distance that is the same or less distance than on the job; or within a job-related amount of time that is the same or less than the amount of time in which the task must be performed on the job).

The amount of time a test taker uses to perform the test events should be carefully measured and recorded. To increase the reliability of time measurements, it is recommended that two administrators should time the test events whenever possible. The use of timers, where the candidate presses a button to begin the test event and presses the same button when they have completed the event, is also helpful.

Administrators should ensure that test results and information are recorded on the appropriate form(s). To minimize potential conflict later, it might help if the scoring form is signed by both the test administrator and the job candidate at the end of testing. However, this is not generally required for a testing process to be considered valid.

The Americans with Disability Act (ADA) and Physical Ability Testing

As mentioned previously, testing that measures the test takers' physiological signs (such as heart or breathing rate) would be considered "medically-based" testing under the ADA. Medically-based tests can only be given after a bona-fide offer of employment has been extended to the job candidate. Furthermore, the ADA specifies that applicants be required to perform the "essential" work functions with or without reasonable accommodations, that these be clearly described to applicants prior to job entry, and may be represented and measured on pre-employment tests. The EEOC indicates that “essential functions are the basic job duties that an employee must be able to perform, with or without reasonable accommodation.” A job function may be considered essential for any of the several reasons, including but not limited to the following:

- The function may be essential because the reason the position exists is to perform the function;
- The function may be essential because of the limited number of employees available among whom the performance of that job function can be distributed; and/or
- The function may be highly specialized so that the incumbent in the position is hired for their expertise or ability to perform the particular function.

Evidence of whether a particular function is essential includes, but is not limited to:

- The employer’s judgment as to which function is essential;
- Written job descriptions prepared before the advertising or interviewing applicants for a job;
- The amount of time spent on the job performing the function;
- The consequences of not requiring the incumbent to perform that function;
- The terms of a collective bargaining agreement;
- The work experience of past incumbents in the job; and/or
- The current work experience of incumbents in similar jobs.[38]

While it may seem counterintuitive, an employer must provide reasonable accommodation to an applicant with a disability during testing *even if* that same employer knows they will be unable to provide this individual with a reasonable accommodation on the job (due to "undue hardship" which the employer must be prepared to prove). The EEOC warns employers to assess the need for accommodations for the application process separately from those that may be needed to perform the job.[39]

The next section provides an example of a physically-challenging test event that has been successfully used for determining if firefighter job candidates can perform one part of the job of a firefighter if they were hired. This shows the level of detail that is advised for developing and administering physical ability testing that would most likely survive a challenge.

Sample Test Event Description: Ladder Removal/Carry

Description: Candidate removes a 24-foot aluminum extension ladder from mounted hooks, carries the ladder a minimum of a total of 60 feet (around a diamond shaped course, the boundaries of which are marked on the ground for them to follow), and replaces the ladder on to the same mounted hooks within three minutes.

Specifications:

- The 24-foot aluminum extension ladder should weigh 41 pounds.
- The mounted hooks should be positioned so that the top portion of the ladder is located 48 inches from the ground.

Demonstration to Candidates: Information that should be given to candidates during the demonstration:

There are three methods that may be used when completing this event: (1) the High Shoulder Carry, the (2) Low Shoulder Carry, and (3) the Suitcase Carry. With all methods, candidates should begin by finding the balance point of the ladder. Rungs in the middle of the ladder, which should provide the best balance point, will be marked.

1. High Shoulder Carry: In the high shoulder carry, the entire ladder sits on the top of the candidate's shoulder. Candidates may place the ladder directly on their shoulder from the mounted hooks and proceed around the designated area, replacing the ladder to the hooks directly from the shoulder.

2. Low Shoulder Carry: In the low shoulder carry, the top beam of the ladder sits on the top of the candidate's shoulder. Candidates may place the ladder directly on their shoulder from the mounted hooks and proceed around the designated area, replacing the ladder to the hooks directly from the shoulder.

3. Suitcase Carry: In this method, the top beam of the ladder is held in one arm like a suitcase.

If, in the administrator's opinion, the candidate loses control of the ladder while carrying it around the designated area, the administrator may intervene. The administrator will take the ladder from the candidate, placing it on the ground at the place where the test taker lost control. The candidate can then pick the ladder up (in any fashion) and continue.

When replacing the ladder, both ends of the ladder must be in control of the test taker and off of the ground.

The ladder must be replaced on the hooks in the original position. There will be rungs painted on the ladder to assist candidates in this process. If the ladder is not replaced in the original position, candidates will be required to remove the ladder and replace it in the proper position.

Scoring: While performing this event, candidates are allowed two penalties before failing. A penalty should be given for any of the following:

- If the candidate drops the ladder, or if it touches the ground;
- If the candidate loses control of the ladder and the administrator must step in and assist;
- If the candidate must place the ladder on the ground to gain stability;
- If the ladder falls over the neck of the candidate, with the candidate's neck between ladder rungs. (In this case, the proctor should immediately assist in the removal and grounding of the ladder);
- If the candidate steps outside of the marked boundary path; and lastly,
- If the candidate fails to follow instructions when performing the test event.
    - If the candidate fails to follow instructions when performing a test event, immediately stop the testing and timing of the event. Explain again how the event should be performed and, if appropriate, provide an additional demonstration. Ask the candidate to acknowledge that they understand the instructions on how to properly perform the test event before allowing the candidate to continue (or start again, depending on the circumstances). This should be allowed twice. If the candidate still does not follow the instructions after testing

begins a third time, the candidate automatically fails this event and testing for that candidate should be discontinued.

Pass/Fail Criteria: The candidate will be automatically disqualified (*i.e.,* fail) if a third penalty occurs or if the event is not successfully completed within three (3) minutes.

Candidates should also be evaluated for their ability to work safely and/or follow safe working practices during the physical ability testing process. The administrator should carefully document any of the following:

- If candidates fail to follow the safety rules and/or procedures of which they had been made aware;
- If candidates ignore potential safety hazards that should have been obvious to a minimally-qualified, entry-level employee the first day on the job prior to training; or
- If, during the event, candidates perform in a way that is in violation of safety protocols that should have been obvious to a minimally-qualified, entry-level employee the first day on the job prior to training.

Methods shown to candidates:

- Safe lifting techniques (*e.g.,* bend knees when lifting)
- Finding the balance point of the ladder
- Placing ladder directly on shoulder from mounted hooks
- High shoulder carry
- Low shoulder carry
- Suitcase carry
- Properly replacing the ladder (with both ends off of the ground and in the appropriate position)

Chapter 7 – Using Physical Ability Tests as "Maintenance Standards" for Incumbent Firefighters

In 2000, the National Fire Protection Agency made a bold but profound statement: "Overweight, out-of-shape fire fighters are an accident waiting to happen" (NFPA, 2000). While the statement can be supported by common sense alone, research data shows just how true this statement really is. For example, a 2005 study revealed that nearly 50% of all injuries to civilian firefighters in that year were a result of sprains, strains, and muscular pain—whereby overexertion is considered the primary causative factor (NIST, 2005).

Firefighters are charged with the serious responsibility of ensuring the safety of their crew and the public. Fire departments are motivated to reduce worker compensation claims, thereby reducing employment costs, which only constitutes some of the costs related to firefighter injuries. After tallying all of the costs related to firefighter injuries in 2002, NIST estimates the annual price to fall between $2.8 and $7.8 billion (NIST, 2005).

This background shows why many fire department executives are passionate about ensuring the high fitness levels of their active fire suppression personnel. While this may be the case, a national research survey of 185 chief-level fire officers[40] revealed that only 25% of fire departments use PATs as *annual maintenance standards* for ensuring the fitness levels of their incumbent fire suppression personnel. This survey revealed that a much higher percentage (88%) use PATs for pre-screening firefighters. So, while fire departments seem intent on screening fit candidates *into* their departments, maintenance testing programs are not typically put into place to continually ensure the fitness level of incumbent fire personnel.

This is not because fire departments do not believe in the importance of ongoing testing. Indeed, this same survey revealed that 93% of the fire chiefs believed that, "Active Fire Suppression Personnel should be tested annually to ensure that they possess the minimum physical abilities necessary to successfully perform the job." This shows overwhelming support for using PATs as a maintenance standard. So, why is there such a gap between this 93% endorsement and the fact that only 25% of fire departments actually use PATs for maintenance standards? Is it the union? Fear of employment lawsuits from personnel who cannot pass the PAT test standard? The answers likely differ from department to department.

Regardless of the reasons behind why the majority (75%) of fire departments do not use a maintenance standard, the reasons for installing a PAT as a maintenance standard are worth serious consideration. In addition to the injuries, the costs from injuries, and the importance of protecting and preserving life and property, there is the fact that firefighters simply *age* after they start the job, and aging has a direct impact on fitness levels. For example, one study[41] involving 256 incumbent fire suppression personnel (with an average age of 34.83 years) revealed a very high correlation (r = .397) between age and test scores (in seconds) on a work sample PAT. This correlation translates to roughly *five seconds slower per year*.

To put this into perspective, a 25 year-old firefighter has a predicted score on the work sample PAT of about eight minutes, whereas a 50 year-old firefighter has a predicted score of ten minutes. This two-minute score difference is attributable to age alone. This trend clearly indicates that age, if left to its natural process without fitness training interventions, will gradually move a minimally-qualified firefighter who (at age 25) barely passed the job-related minimum cutoff score (9 minutes and 34 seconds on this particular PAT), to a score that is one full minute slower in just 12 years.

This phenomenon presents fire departments with three options: (1) do nothing and cope with a workforce with naturally declining physical abilities, (2) install a wellness program and hope that job-related standards associated with important fire suppression tasks are positively impacted, or (3) install a wellness program *coupled with an annual maintenance standard using a work sample PAT*. The latter option actually ensures that active fire suppression personnel will maintain job performance standards.

Departments that adopt work sample PATs as an annual maintenance standard must address three controversial issues: (1) selecting an appropriate cutoff time for the test (the same time used for entry level or slower/faster), (2) choosing which positions will be selected for the annual testing requirement, and (3) identifying the steps that will be taken with incumbents who cannot pass the annual test, even after repeated retest opportunities. These issues are addressed next.

Selecting a Cutoff Time for a Maintenance Standard Work Sample PAT

The process for setting an entry-level (pre-hire) cutoff for a work sample PAT described in the previous chapter (*i.e.*, averaging the minimum passing time recommendations from Job Experts) will not typically work effectively for establishing an annual maintenance cutoff standard. This is for the simple reason that firefighter ratings (cutoff opinion times) will likely be biased (on the slow side) because they are being asked to set a standard that may possibly result in *one of their teammates losing their jobs.*

For this reason, the recommended procedure for determining a maintenance standard time cutoff is to run a representative sample of active fire suppression personnel through the test and use their *observed scores* in a process to compute the cutoff score. While the incumbents can still bias the time cutoff established using this process (by completing the test slower than their true ability levels would allow), proprietary studies[42] have shown that using firefighter actual times will result in more accurate cutoffs than using their opinion times (at least when it comes to setting maintenance standards).

The process involved in setting the maintenance standard cutoff is described below:

1. Convene a representative sample of at least 30 fire incumbents from the active fire suppression positions in your department (see additional guidelines for this below). While 30 incumbents should serve as a minimum, a larger sample is more desirable.

2. Have the incumbents complete the test and record their actual times for completing the PAT.

3. Trim the outliers from the study by removing test scores that are 1.645 standard deviations above or below the average.

4. Obtain a Standard Error of Measurement (SEM) for the PAT by conducting a test-retest study. This requires having 60+ applicants or incumbents taking the PAT twice (separated by 1-2 weeks), and correlating the two scores to obtain the test-retest reliability estimate (r_{tt}). This value is used along with the standard deviation of the sample to compute the SEM using the formula: $SEM = \sigma_x * \sqrt{(1 - r_{tt})}$ where σ_x is the standard deviation of test scores (*e.g.*, PAT times in seconds) and r_{tt} is the test-retest reliability. For example, if the test-retest reliability is .84 and the standard deviation is 10, the SEM would be 4.0 (in Excel: = 10*(sqrt(1-.84)). As an example, the SEM for the FPSI Work Sample PAT is r = .8015.

5. Compute the Standard Error of Difference (SED) using the formula: $SEM * \sqrt{2}$.

6. Compute the Standard Error of the Mean (SE_{Mean}) using the formula: $SE_{Mean} = \sigma_x * \sqrt{N}$ where σ_x is the standard deviation of test scores (*e.g.*, PAT times in seconds) and *N* is the number of fire incumbents who completed the test. For example, if the standard deviation is 30 and 45 fire incumbents completed the test, the SE_{Mean} would be 4.47 (in Excel: = 30/(sqrt(45)). Using the Standard Error of the Mean (SE_{Mean}) in the formula below to set the final incumbent cutoff will offset some of the sampling reliability concerns that may

emerge from testing less than all active fire suppression personnel in the target department.

7. Compute the final cutoff score using the formula:

   Final cutoff = (Trimmed) Mean + SE_{Mean} + (1.645*SED).

This process sets the cutoff at a level that represents the minimum standard needed for active fire suppression personnel. Incumbents who take longer than this time limit fall outside of the minimum proficiency levels relating to physical ability requirements for the job (*i.e.*, with a 95% confidence level that these scores are reliably different from the average parts of the score distribution), and should be required to improve their abilities through possible dietary changes, weight-loss programs, and/or physical fitness programs.

Which Positions Should be Included in an Annual Maintenance Testing Program?

When fire chiefs who participated in the research study were asked the controversial question regarding which ranks should be required to pass an annual maintenance PAT, the results showed a clear cluster that included four ranks: Firefighter, Fire Engineer, Fire Lieutenant, and Fire Captain. Over 70% of the survey respondents were in clear agreement that maintenance PATs would be appropriately required for these positions. The next cluster included the Training Officer and Battalion Chief positions, which were both tied at about 60% agreement. The higher-level ranks (which included Fire Marshall, Division Chief, Assistant Chief, and Chief) fell between 30% and 40%, indicating that being able to pass an annual maintenance PAT was clearly less important for these ranks. Figure 2 shows these results graphically.

Figure 2. Fire Personnel Required to Pass Annual Maintenance PATs

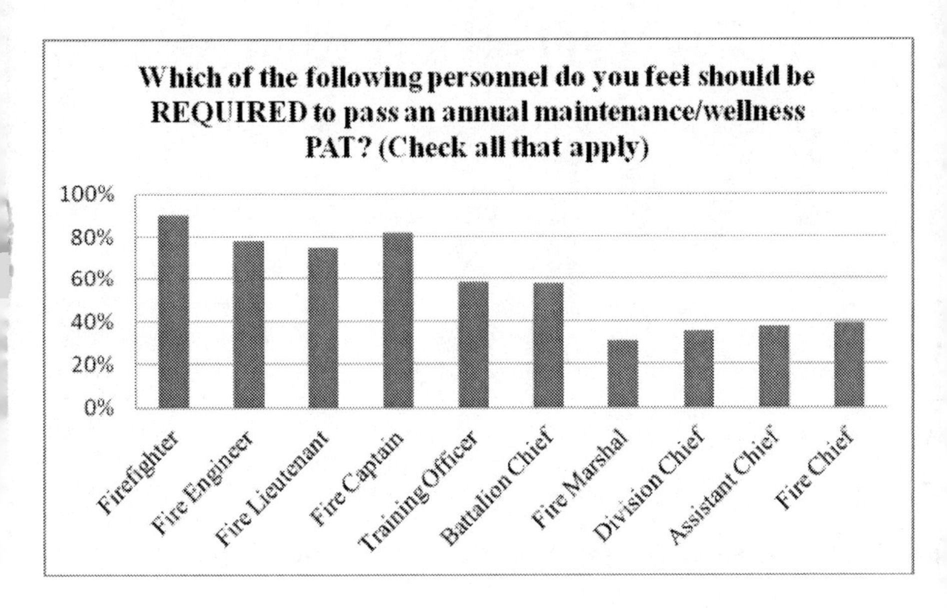

Figure 3 below shows the percentage of time that various ranks spend in active fire suppression activities. These results reveal the reasons behind the results provided in Figure 2—*i.e.*, the importance of using a maintenance PAT is directly tied to the percentage of time that various ranks spend in fire suppression activities.

Figure 3. Percentage of Time Spent in Active Fire Suppression Activities (by rank)

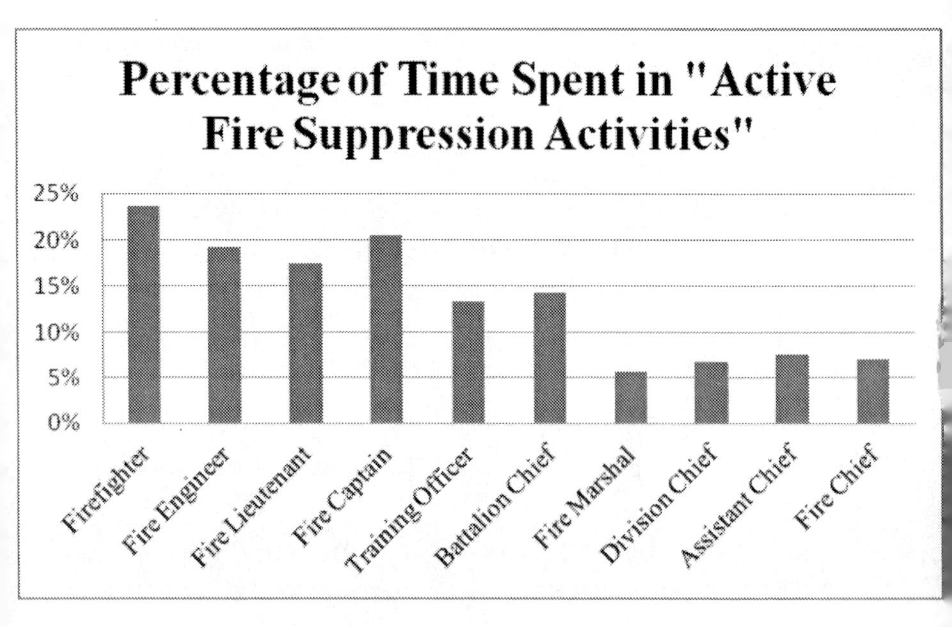

The study revealed that the average percentage of calls that were fire suppression calls was 21%, with a standard deviation of 13.5%. The percentage of calls that were EMS was 71%, with a standard deviation of 14.8%. There was no correlation between department size and type of calls, which reveals that the ratio of fire to EMS calls is not dependent on department size.

Choosing which positions to include in an annual maintenance testing program should clearly be a department-by-department decision. With that said, the data reveal that the four positions that are traditionally "hands on" when it comes to fire scene management should certainly be included in most situations. This includes the ranks of Firefighter, Fire Engineer, Fire Lieutenant, and Fire Captain. In most departments, the

Training Officer is not directly involved in responding to fire emergencies. The Battalion Chief position, however, is different because field deployment levels of this position is sometimes high, and will vary by assignment (e.g., training, administrative, etc.) as well as department size. The higher-level ranks (*e.g.*, Fire Marshall, Division Chief, Assistant Chief, and Chief) will typically be exempt from maintenance programs.

What Steps Should Departments Take with Incumbents who Fail Annual Maintenance Standards?

The research conducted surrounding this issue included a question that asked respondents: "Which of the following consequences do you feel are acceptable for ACTIVE FIRE SUPPRESSION who cannot pass a maintenance/wellness PAT?" The four response options that were provided to respondents were:

- Conditioning program—The incumbent is placed on a program that includes dietary modification and physical training.
- Leave of absence—The department may elect to place the incumbent on a leave of absence until which time the incumbent is able to pass the test.
- Disability leave—The department may elect to place the incumbent on disability leave until which time the incumbent is able to pass the test.
- Retirement with pension—The department may elect to terminate employment with the incumbent following continued attempts to improve test performance without success.

The results from this survey question are provided in Figure 4.

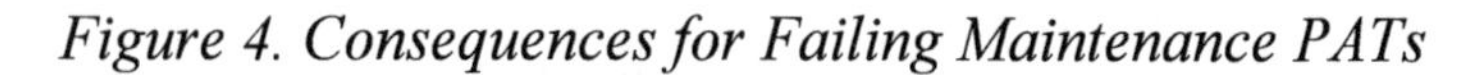

Figure 4. Consequences for Failing Maintenance PATs

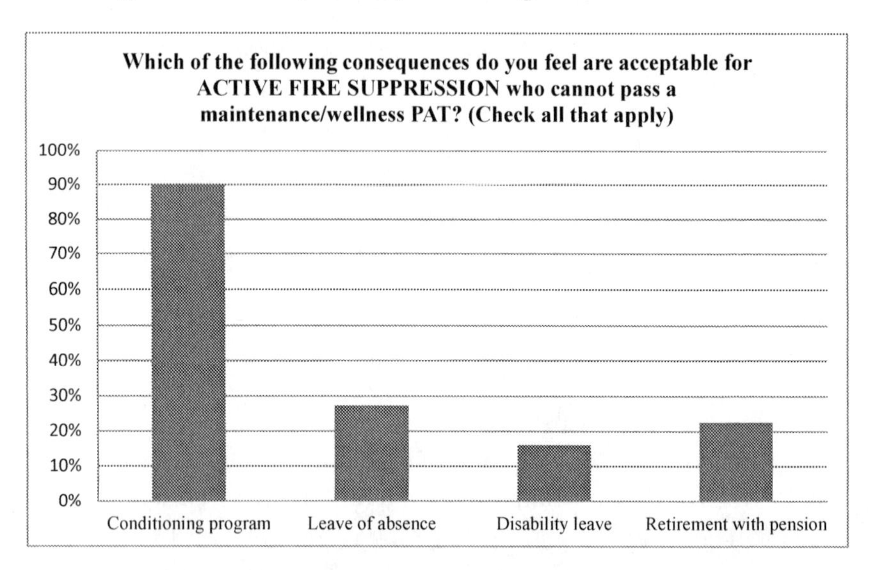

Figure 4 shows that most responding chiefs (90%) agreed that requiring a condition program was a sound "natural consequence" for incumbents who cannot pass a maintenance PAT. However, a significant portion of the chiefs stated that more severe consequences (taking a leave of absence or retirement with pension) would also be justified (with 27% and 22% endorsement, respectively). Only 15% endorsed the most extreme consequence (required disability leave).

Before moving to one of these three severe consequences, we suggest first allowing the candidate (up to) two retesting opportunities (each separated by a 10-16 week training program). The 10-16 week training program should consist of both cardiovascular and strength training in the specific, fire suppression-related work behaviors that are measured by the test. Departments can choose whether they want the training program to be self-directed or conducted by a department-designated exercise specialist.

Chapter 8 – Would Your Tests Survive a Legal Challenge? Checklists for Evaluating Your Test's Validity

Most tests in the fire service are supported under a content validation model. Some tests, such as personality tests and some types of cognitive ability tests, are supported using criterion-related validity. There are fundamental requirements under the *Uniform Guidelines* that should be addressed when a department claims either type in an enforcement/litigation setting. These are provided in Tables 7-11.

Table 7. Content Validation Checklist for Written Tests

Req. #	Uniform Guidelines Requirement	Uniform Guidelines Reference
1	Does the test have **sufficiently high reliability?** (Generally, written tests should have reliability values that **exceed .70**[1] **for each section of the test that applicants are required to pass**).	14C(5)
2	Does the test measure Knowledges, Skills, or Abilities (KSAs) that have been rated as **"critical" (necessary prerequisites** for performance of the job) for each job that used the test? (Or, does the test measure KSAs that are "important" which are also clearly linked to critical (necessary for performance of the job) job duties, or job duties that constitute most of the job.)	14C(1,4,8)
3	If the test is used for testing the KSAs of **higher-level positions**, are job progression structures so established that employees will probably, within a reasonable period of time and in a majority of cases, progress to those higher level jobs?	5I
4	Does the test measure KSAs that are **necessary on the first day of the job**? (Check "No" if the KSAs measured by the test will be trained on the job or can be "learned in a brief orientation").	14C(1), 5F, 5I(3)
5	Does the test **measure KSAs that are concrete and not theoretical?** (Under content validity, tests cannot measure abstract "traits" such as intelligence, aptitude, personality, common sense, judgment, leadership, or spatial ability, if they are not defined in concrete, observable ways).	14C(1,4)
6	**Is sufficient time allowed for nearly all applicants to complete the test?** [2] (Unless the test was specifically validated with a time limit, sufficient time should be allowed for nearly all applicants to finish).	15C(5)
7	FOR TESTS MEASURING JOB KNOWLEDGE ONLY: Does the test measure job knowledge areas that need to be **committed to memory?** (Check "No" if the job knowledge areas can be easily looked up without hindering job performance).	15C(3), Q&A 79
8	Were alternative procedures that are "substantially equally valid," but have less adverse impact, investigated?	3B, 15B(9)

Notes: (1) U.S. Department of Labor, Employment and Training Administration (2000). Testing and Assessment: An Employer's Guide to Good Practices. (2) Crocker, L. & Algina, J. (1986). *Introduction to classical and modern test theory* (p. 145). Fort Worth, TX: Harcourt Brace Jovanovich.

Table 8. Criterion-related Validation Checklist for Written Tests

Req. #	Uniform Guidelines Requirement	Uniform Guidelines Reference
1	Is there a description of the test? Look for title, description, purpose, target population, administration, scoring and interpretation of scores.	15B(4)
2	If the test is a combination of other tests or if the final score is derived by weighting different parts of the test or different tests, is there a description of the rationale and justification for such combination or weighting?	15B(10)
3	Does the test have sufficiently high reliability (e.g., .70[1] is desirable).	15C(7)
4	Is there a description of the criterion measure, including the basis for its selection or development and method of collection? For ratings, look for information related to the rating form and instructions to raters.	15B(5)
5	Does the criterion (*i.e.*, performance) measure reflect either: (a) important or critical work behaviors or outcomes as identified through a job analysis or review or (b) an important business need, such as absenteeism, productivity, tardiness or other?	14B(2)(3), 15B(3)
6	Is the sample size adequate **for each position that validity is being claimed**? Look for evidence that the correlations between the predictor and criterion measures are sufficient for each position included in the study.	14B(1)
7	Is the study sample representative of all possible test-takers? Look for evidence that the sample was chosen to include individuals of different races and gender. For concurrent validity studies, look for evidence that the sample included individuals with different amounts of experience. Where a number of jobs are studied together (e.g., a job group), look for evidence that the sample included individuals from all jobs included in the study.	14B(4)
8	Are the methods of analysis and results described? Look for a description of the method of analysis, measures of central tendency such as average scores, measures of the relationship between the predictor and criterion measures and race/gender breakouts.	15B(8)
9	Is the correlation between scores on the test and the criterion statistically significant **before** applying any statistical corrections?	14B(5)

Req. #	Uniform Guidelines Requirement	Uniform Guidelines Reference
10	Is the test being used for the same jobs for which it was validated? For the same type of test-takers?	14B(6), 15B(10)
11	Have steps been taken to correct for overstatement and understatement of validity findings, such as corrections for range restriction, use of large sample sizes or cross-validation? If corrections are made, are the raw and corrected values reported?	14B(7)
12	Has the fairness of the test been examined or, if not feasible, is there a plan to conduct such a study?	14B(8)
13	Has a validation study been conducted in the last 5 years or, if not, is there evidence that the job has not changed since the last validity study?	5K
14	Were alternative procedures that are "substantially equally valid," but have less adverse impact, investigated?	3B, 15B(9)
15	**If criterion-related validity for the test is being "transported"** from another employer/position, were the following requirements addressed? (a) the original validation study addressed Section 14B of the Guidelines, (b) the jobs perform substantially the same major work behaviors (as shown by job analyses in both locations), (c) a fairness study was conducted (if technically feasible).	7B, 14B

Notes: (1) U.S. Department of Labor, Employment and Training Administration (2000). Testing and Assessment: An Employer's Guide to Good Practices.

Table 9. Content Validation Checklist for Interviews

Req. #	Uniform Guidelines Requirement	Uniform Guidelines Reference
1	IF MULTIPLE RATERS ARE INVOLVED IN THE INTERVIEW ADMINISTRATION/SCORING[1], does the interview have **sufficiently high inter-rater reliability?** (Generally, interviews should have reliability values that **exceed .60**[2] for each section of the interview that applicants are required to pass).	14C(5)
2	Does the interview measure Knowledges, Skills, or Abilities (KSAs) that have been rated as **"critical" (necessary prerequisites** for performance of the job) for each job that used the test? (Or, does the test measure KSAs that are "important" which are also clearly linked to critical (necessary for performance of the job) job duties, or job duties that constitute most of the job.)	14C(1,4,8)
3	If the interview is used for testing the KSAs of **higher-level positions**, are job progression structures so established that employees will probably, within a reasonable period of time and in a majority of cases, progress to those higher level jobs?	5I
4	Does the interview measure KSAs that are **necessary on the first day of the job**? (Check "No" if the KSAs measured by the interview will be trained on the job or can be "learned in a brief orientation").	14C(1), 5F, 5I(3)
5	Does the interview **measure KSAs that are concrete and not theoretical?** (Under content validity, tests cannot measure abstract "traits" such as intelligence, aptitude, personality, common sense, judgment, leadership, or spatial ability, if they are not defined in concrete, observable ways).	14C(1,4)
6	FOR INTERVIEWS MEASURING JOB KNOWLEDGE ONLY: Does the interview measure job knowledge areas that need to be **committed to memory?** (Check "No" if the job knowledge areas can be easily looked up without hindering job performance).	15C(3), Q&A 79
7	Were alternative procedures that are "substantially equally valid," but have less adverse impact, investigated?	3B, 15B(9)

Notes: (1) "Internal consistency" reliability for interviews with one or multiple raters is also important; however, inter-rated reliability sets the upper limit on reliability when multiple-rater panels are used. (2) Most resources recommend using reliability levels of at least .70 as a baseline (e.g., U.S. Department of Labor, Employment and Training Administration (2000). Testing and Assessment: An Employer's Guide to Good Practices), however, interviews typically have lower reliability levels than written tests.

Table 10. Content Validation Checklist for Work Sample (WS) or Physical Ability Tests (PATs)

Req. #	Uniform Guidelines Requirement	Uniform Guidelines Reference
1	Does the WS/PAT have **sufficiently high reliability**? (Typically, WS/PATs need to be supported using test-retest reliability, unless they have a sufficient number of scored components to be evaluated using internal consistency. Generally, WS/PATs should have reliability values that exceed **.70**(1) for each section of the test that applicants are required to pass).	14C(5)
2	Does the WS/PAT measure Knowledges, Skills, or Abilities (KSAs) that have been rated as **"critical" (necessary prerequisites** for performance of the job) for each job that used the test? (Or, does the test measure KSAs that are "important" which are also clearly linked to critical (necessary for performance of the job) job duties, or job duties that constitute most of the job.)	14C(1,4,8)
3	If the WS/PAT is used for testing the KSAs of **higher-level positions**, are job progression structures so established that employees will probably, within a reasonable period of time and in a majority of cases, progress to those higher level jobs?	5I
4	Does the WS/PAT measure KSAs that are **necessary on the first day of the job**? (Check "No" if the KSAs measured by the WS/PAT will be trained on the job or can be "learned in a brief orientation").	14C(1), 5F, 5I(3)
5	Does the WS/PAT measure KSAs that are **concrete and not theoretical**? (Under content validity, tests cannot measure abstract "traits" such as intelligence, aptitude, personality, common sense, judgment, leadership, or spatial ability, if they are not defined in concrete, observable ways). Measuring "general strength," "fitness," or "stamina" cannot be supported under content validity unless they are *operationally defined in terms of observable aspects of work behavior* (job duties).	14C(1,4), 15C(5)
6	If the WS/PAT is designed to replicate/simulate actual work behaviors, is the **manner, setting, and level of complexity** highly similar to the job?	14C(4)
7	**If the WS/PAT has multiple events and is scored using a time limit** (e.g., all events must be completed in 5 minutes or faster), are the events in the WS/PAT typically performed on the job **with other physically-demanding duties performed immediately prior to and after each event**?	15C(5)

Req. #	Uniform Guidelines Requirement	Uniform Guidelines Reference
8	**If the WS/PAT has multiple events and is scored using a time limit** (e.g., all events must be completed in 5 minutes or faster), is **speed** typically important when these duties are performed on the job?	15C(5)
9	**If the WS/PAT includes weight handling requirements** (e.g., lifting, carrying certain objects or equipment), do they represent the **weights, distances, and duration** that objects/equipment are typically carried by a **single person** on the job?	15C(5)
10	If there are any **special techniques** that are learned on the job that allow current job incumbents to perform the events in the test better than an applicant could, are they demonstrated to the applicants before the test?	14C(1), 5F, 5I(3)
11	Does the WS/PAT **require the same or less exertion** of the applicant than is required on the job?	5H, 15C(5)
12	Were alternative procedures that are "substantially equally valid," but have less adverse impact, investigated?	3B, 15B(9)

Notes: (1) U.S. Department of Labor, Employment and Training Administration (2000). Testing and Assessment: An Employer's Guide to Good Practices.

Table 11. Validation Checklist for Using Test Results

Req. #	Uniform Guidelines Requirement	Uniform Guidelines Reference
1	If a **pass/fail cutoff** is used, is the cutoff "set so as to be reasonable and consistent with normal expectations of acceptable proficiency within the work force"?	5G, 5H, 15C(7)
2	If the test is **ranked or banded above a minimum cutoff level, and is based on content validity**, can it be shown that either **(a)** applicants scoring below a certain level have little or no chance of being selected for employment, or **(b)** the test measures KSAs / job duties that are "performance differentiating"?[1]	3B, 5G, 5H, 14C(9)
3	If the test is **ranked or banded above a minimum cutoff level, and is based on criterion-related validity**, can it be shown that either **(a)** applicants scoring below a certain level have little or no chance of being selected for employment, or **(b)** the degree of statistical correlation and the importance and number of aspects of job performance covered by the criteria clearly justify ranking rather than using the test in a way that would lower adverse impact (*e.g.,* banding or using a cutoff)? (Tests that have adverse impact and are used to rank that are only related to one of many job duties or aspects of job performance should be subjected to close review.)	3B, 5G, 5H, 14B(6)
4	Is the test **used**[2] **in a way that minimizes adverse impact**? (Options include different cutoff points, banding, or weighting the results in ways that are still "substantially equally valid" but reduce or eliminate adverse impact)?	3B, 5G

Notes: (1) "Performance Differentiating" KSAs are those that differentiate between "adequate" and "superior" job performance (see UGESP, Section 14C[9]). (2) "Alternate uses" of a practice, procedure, or test can include different cutoff points, banding, or weighting the results in ways that are still "substantially equally valid."

References

Biddle, D. A. (2011). *Adverse Impact and Test Validation: A Practitioner's Handbook* (3rd ed.). Scottsdale, AZ: Infinity Publishing.

Biddle, D. A. & Morris, S. B. (2011). Using Lancaster's mid-P correction to the Fisher's Exact Test for adverse impact analyses. *Journal of Applied Psychology*, 96 (5), 956-965.

Biddle Consulting Group, Inc. (2011). Test Validation & Analysis Program (TVAP, Version 7) [Computer Software]. Folsom, CA: Author.

Boston Chapter, NAACP, Inc. v. Beecher, 504 F.2d 1017, 1026-27 (1st Cir. 1974).

Brunet v. City of Columbus, 1 F.3d 390, C.A.6 (Ohio, 1993).

Clady v. County of Los Angeles, 770 F.2d 1421, 1428 (9th Cir., 1985).

Dye, D. A., Reck, M., & McDaniel, M. A. (1993, July). The validity of job knowledge measures. *International Journal of Selection and Assessment, 1* (3), 153-157.

Griggs v. Duke Power, 401 U.S. 424 (1971).

Hunter, J.E. (1983). A causal analysis of cognitive ability, job knowledge, job performance and supervisory ratings. In F. Landy and S. Zedeck and J. Cleveland (Eds.), *Performance measurement theory* (pp. 257-266). Hillsdale, NJ: Erlbaum.

National Fire Protection Agency (2000). *NFPA 1583: Standard on Health-Related Fitness Programs for Fire Fighters*. Quincy, Massachusetts: Author.

National Institute of Standards and Technology (NIST) (March, 2005). *The Economic Consequences of Firefighter Injuries and Their Prevention. Final Report*. Arlington, VA: Author.

Ricci v. DeStefano, 129 S. Ct. 2658, 2671, 174 L. Ed. 2d 490 (2009).

Sackett, P. R., Schmitt, N., Ellingson, J. E., & Kabin, M. B. (2001). High-stakes testing in employment, credentialing, and higher education: Prospects in a post-affirmative-action world. *American Psychologist, 56* (4), 302-318

Siconolfi, S. F., Garber, C. E., Lasater, T. M., & Carleton, R. A. (1985). A simple, valid step test for estimating maximal oxygen uptake in epidemiologic studies. *American Journal of Epidemiology, 121*, 382-390.

SIOP (Society for Industrial and Organizational Psychology, Inc.) (1987, 2003), *Principles for the Validation and Use of Personnel Selection Procedures* (3rd and 4th eds). College Park, MD: SIOP.

Stagi v. National Railroad Passenger Corporation, No. 09-3512, 3d Cir. (Aug. 16, 2010).

National institute of Standards and Technology (NIST) (March, 2005). *The Economic Consequences of Firefighter Injuries and Their Prevention. Final Report*. Arlington, VA: Author.

Uniform Guidelines – Equal Employment Opportunity Commission, Civil Service Commission, Department of Labor, and Department of Justice (August 25, 1978), Adoption of Four Agencies of Uniform Guidelines on Employee Tests, 43 Federal Register, 38,290-38,315; Adoption of Questions and Answers to Clarify and Provide a Common Interpretation of the Uniform Guidelines on Employee Tests, 44 Federal Register 11,996-12,009.

U.S. Department of Labor: Employment and training administration (2000), *Testing and Assessment: An Employer's Guide to Good Practices*. Washington DC: Department of Labor Employment and Training Administration.

Vulcan Society v. City of New York, 07-cv-2067 (NGG)(RLM) (July 22, 2009).

Ward's Cove Packing Co. v. Atonio, 490 U.S. 642 (1989).

Zamlen v. City of Cleveland, 686 F.Supp. 631, N.D. (Ohio, 1988).

Endnotes

---

[1] In this text, disparate impact and adverse impact mean the same.

[2] Readers interested in the historical and theoretical background of adverse impact are encouraged to read Biddle, D. A. (2011). *Adverse Impact and Test Validation: A Practitioner's Handbook* (3rd ed.). Scottsdale, AZ: Infinity Publishing).

[3] See http://www.disparateimpact.com for an online tool for computing adverse impact.

[4] The Fisher Exact Test should not be used without this correction—see Biddle & Morris (2011) (cited in the reference section of this book).

[5] See, for example: OFCCP v. TNT Crust (U.S. DOL, Case No. 2004-OFC-3); Dixon v. Margolis (765 F. Supp. 454, N.D.Ill., 1991), Washington v. Electrical Joint Apprenticeship & Training Committee of Northern Indiana, 845 F.2d 710, 713 (7th Cir.), cert. denied, 488 U.S. 944, 109 S.Ct. 371, 102 L.Ed.2d 360 (1988). Stagi v. National Railroad Passenger Corporation, No. 09-3512 (3d Cir. Aug. 16, 2010).

[6] While these guidelines are suitable for most tests that have either a single or a few highly-related abilities being measured, sometimes wider guidelines should be adopted for multi-faceted tests that measure a wider range of competency areas (*e.g.*, situational judgment, personality, behavior, bio-data tests).

[7] The survey was limited to competency areas that can be possibly measured in a testing process.

[8] When tests are based on criterion-related validity studies, cutoffs can be calibrated and set based on empirical data and statistical projections that can also be very effective.

[9]For example US v. South Carolina (434 US 1026, 1978) and Bouman v. Block (940 F.2d 1211, C.A.9 Cal., 1991) and related consent decrees.

[10] Be careful to first remove unreliable and outlier raters before averaging item ratings into a cutoff score. One way to remove outliers is to eliminate ratings (not raters, but only their ratings that have been identified as outliers) using a 1.645 standard deviation rule (*i.e.*, all ratings that are 1.645 standard deviations above or below the mean are deleted). This process will "trim" the average ratings that are in the upper or lower 5% of the distribution.

[11] See the Biddle (2011) for recommended strategies for computing the C-SEM, SED, and creating score bands.

[12] The SED should also be set using a conditional process wherever feasible (see Biddle, 2011). Multiplying the SED by a confidence interval, such as 1.645 (95%, one directional) is also an acceptable practice.
[13] For example, Schmidt, F. L. (1991), 'Why all banding procedures in personnel selection are logically flawed', *Human Performance*, 4, 265-278; Zedeck, S., Outtz, J., Cascio, W. F., and Goldstein, I. L. (1991), 'Why do "testing experts" have such limited vision?', *Human Performance*, 4, 297-308.
[14] One clear reason for using banding as a means of reducing adverse impact can be found in Section 3B of the Uniform Guidelines which states: "Where two or more tests are available which serve the user's legitimate interest in efficient and trustworthy workmanship, and which are *substantially equally valid* for a given purpose, the user should use the procedure which has been demonstrated to have the lesser adverse impact." Banding is one way of evaluating an alternate use of a test (*i.e.,* one band over another) that is "substantially equally valid."
[15] See, for example: Officers for Justice v. Civil Service Commission (CA9, 1992, 979 F.2d 721, cert. denied, 61 U.S.L.W. 3667, 113 S. Ct. 1645, March 29, 1993).
[16] See Section 14C9 of the Uniform Guidelines.
[17] For example, in Guardians v. CSC of New York (630 F.2d 79), one of the court's reasons for scrutinizing the use of rank ordering on a test was because 8,928 candidates (two-thirds of the entire testing population) was bunched between scores of 94 and 97 on the written test.
[18] See, for example, Gatewood, R. D. & Feild, H.S. (1994), *Human Resource Selection* (3rd ed.) Fort Worth, TX: The Dryden Press (p. 184); Aiken, L.R. (1988). *Psychological Testing and Assessment* (2nd ed.). Boston: Allyn & Bacon (p. 100); Weiner, E. A. & Stewart, B. J. (1984). *Assessing Individuals.* Boston: Little, Brown. (p. 69).
[19] For tests that are designed to directly mirror job duties (called "work sample tests"), only test-duty (and not test-KSAPC) linkages are required for a content validity study (see Section 14C4 of the Guidelines). In this case, the Best Worker ratings on the duties linked to the work sample test should be the primary consideration for evaluating its use (*i.e.,* ranking or pass/fail). For tests measuring KSAPCs (and not claiming to be direct "work sample tests"), the extent to which the test measures KSAPCs that are differentiating should be the primary consideration.

[20] See, for example, Mosier, C.I. (1943). On the reliability of a weighted composite. *Psychometrika, 8*, 161-168. (6,11).

[21] See Uniform Guidelines Questions & Answers #47, the *Principles* (2003, p. 20, 47), and Cascio, W. (1998), *Applied Psychology in Human Resource Management,* Upper Saddle River, NJ: Prentice-Hall, Inc. for more information on this approach.

[22] For a detailed description of this process, see: Feldt, L.S., & Brennan, R.L. (1989), *Reliability.* In R.L. Linn (Ed.), Educational Measurement (3rd ed.). New York, Macmillan. (pp. 105-146).

[23] For example, see Bouman v. Block (940 F2d 1211, 9th Cir 1991); Hearn v. City of Jackson, Miss. (110 Fed. 242, 5th Cir 2004); Isabel v. City of Memphis (F.Supp.2d 2003, 6th Cir 2003); Paige v. State of California (102 F.3d 1035, 1040, 9th Cir 1996).

[24] The steps outlined in this section are based on the requirements outlined by the *Uniform Guidelines* (1978), the *Principles* (2003), and the *Standards* (1999). The proposed model is not a one-size-fits-all process, but rather a generic template which could be employed in an ideal setting. While it is not *guaranteed* that, by following these steps, litigation will be avoided, implementing the practices outlined in this section will greatly increase the likelihood of success in the event of a challenge to a written testing process.

[25] The Uniform Guidelines do not require frequency ratings for content validity; however, obtaining frequency ratings provides useful information for addressing the 1990 Americans with Disabilities Act (ADA) and can also help when developing a test using content validity.

[26] This rule-of-thumb time limit is only applicable for conventional multiple-choice tests. Where many calculations are needed for each test item (*e.g.*, hydraulic items on a Fire Engineer test), obtain input from Job Experts to ensure an appropriate time limit.

[27] See, for example: Crocker, L. & Algina, J. (1986). *Introduction to classical and modern test theory* (p. 145). Fort Worth, TX: Harcourt Brace Jovanovich.

[28] See Biddle (2011) or the Test Validation & Analysis Program (TVAP®) for the recommended procedures for this step.

[29] VO2 maximum refers to the highest rate of oxygen consumption attainable during maximal or exhaustive exercise.

[30] The U.S. Equal Employment Opportunity Commission, Notice Number 915.002 (Date 10/10/95). See (http://www.eeoc.gov/policy/docs/preemp.html).

[31] Section 703 of the Civil Rights Act of 1964 (42 U.S.C. 2000e-2) (as amended by section 105) states: "It shall be an unlawful employment practice for a respondent, in connection with the selection or referral of applicants or candidates for employment or promotion, to adjust the scores,

use different cutoff scores for, or otherwise alter the results of, employment related tests on the basis of race, color, religion, sex, or national origin."

[32] Alternatively, 1.96 SEDs can be used, which provides 97.5% confidence.

[33] The classical SED is computed by multiplying the SEM by the square root of 2.

[34] Because incumbent firefighters are typically not as motivated as entry-level candidates when taking a Physical Ability Test, sometimes their test scores (*e.g.*, PAT times) may not accurately represent minimum competency levels needed for the job.

[35] See, for example, *EEOC v. Dial Corp.*, 469 F.3d 735 (8th Cir. 2006). A major company lost a 3.4 million-dollar court case due to the fact that, during testing, job candidates were required to not only perform a work-related task at a faster pace than required to be performed on the job, but also without the short breaks between efforts that are allowed on the job.

[36] This will help to emphasize to the candidate the importance of safety on the job and will also help minimize potential grievances that might claim that the safety rules and/or safe working practices were not explained sufficiently.

[37] Test takers frequently do not accurately recall how well they performed during testing. For this reason, video- and/or audio-recording of test events will often enhance an employer's ability to successfully defend the elimination of an unqualified job candidate. As with all test-related documents, recordings should be retained in the event of challenge by job candidates, which can sometimes occur years after testing has taken place. In addition, recordings of job candidates should only be used for selection purposes.

[38] See the 1990 *Americans with Disabilities Act*, Section 1630.2(n).

[39] See Question #13 in the EEOC's *Enforcement Guidance: Reasonable Accommodation and Undue Hardship Under the Americans with Disabilities Act document* at http://www.eeoc.gov/policy/docs/accommodation.html for more details.

[40] The study was conducted by the authors in 2011. The survey sample included 151 Fire Chiefs, 12 Assistant Fire Chiefs, 8 Battalion Chiefs, 6 Deputy Chiefs, 4 Deputy Fire Chiefs, 4 Division Chiefs (185 total). The average department size was 123, with an average of 109 active fire suppression personnel. The smallest department included had 9 full-time employees; the largest had 1,790.

[41] Study conducted by FPSI (2011) involving firefighter incumbents from over 40 fire departments on a single PAT.

[42] Conducted by the authors for confidential clients.